NORTHERN PIKE

Northern Pike
Ecology, Conservation, and Management History

Rodney B. Pierce

Minnesota Department of
Natural Resources

University of Minnesota Press
Minneapolis • London

Frontispiece, page ii: The northern pike (*Esox lucius* Linnaeus). Photograph courtesy of Bill Lindner Photography, Baxter, Minnesota (www.blpstudio.com).

The University of Minnesota Press gratefully acknowledges financial assistance provided for this book by the Minnesota Department of Natural Resources.

Unless otherwise credited, photographs and illustrations are by the author.

Copyright 2012 by the State of Minnesota, Department of Natural Resources

All rights reserved. No part of this publication may be reproduced, stored in a retrieval system, or transmitted, in any form or by any means, electronic, mechanical, photocopying, recording, or otherwise, without the prior written permission of the publisher.

Published by the University of Minnesota Press
111 Third Avenue South, Suite 290
Minneapolis, MN 55401-2520
http://www.upress.umn.edu

Library of Congress Cataloging-in-Publication Data
Pierce, Rodney B.
  Northern pike: ecology, conservation, and management history / Rodney B. Pierce.
    Includes bibliographical references and index.
    ISBN 978-0-8166-7954-6 (hc)
  1. Pike. 2. Pike fisheries—Minnesota—Management. I. Title.
  QL638.E7P54 2012
  597.5'909776—dc23
                                                                    2012005658

Text design and composition by Chris Long/Mighty Media, Inc.

Printed in China on acid-free paper

The University of Minnesota is an equal-opportunity educator and employer.

20 19 18 17 16 15 14 13 12   10 9 8 7 6 5 4 3 2 1

Contents

Acknowledgments	vii
County Map of Minnesota	ix
Introduction: Minnesota's Northern Pike Legacy	xi
1. The Ecology of Northern Pike	1
2. Recreational and Commercial Fishing	71
3. Conservation and Management of Northern Pike	91
4. Sampling Northern Pike Populations	135
Closing Thoughts about Pike Management in Minnesota	175
Bibliography	183
Index	203

Acknowledgments

This book is dedicated to career employees of the Department of Natural Resources in Minnesota and Wisconsin and of similar agencies elsewhere. The professional people I have encountered in the field of natural resources deserve much more credit than they are often given. Their quality of work and the services they provide in the face of numerous obstacles are more than comparable to any in private industry. In particular, the fisheries staff in Grand Rapids, Minnesota, provided the setting that enabled my work. I thank Gary Phillips, Bruce Vondracek, Paul Cunningham, and Sandy DeLeo for helpful reviews and edits.

Being part of the fisheries research group for the Minnesota Department of Natural Resources has been important to me. The group has a well-deserved reputation throughout the United States as second to none. Old-timers from the group have been mentors, and the newer folks are tackling difficult issues. Their backup during challenging spring field seasons was an enormous boost. Personal supervisors Jack Wingate, Dennis Schupp, Don Olson, Don Pereira, and Melissa Drake gave me the outstanding support needed to accomplish long-term research. I benefited from new energy levels sparked by Andy Carlson over the past two years, and my longtime colleague Cindy Tomcko deserves a huge share of credit for work we accomplished with northern pike. Cindy spent a large part of her career doing the heavy lifting with me, including ice-out trapping in blizzard conditions and enduring many days of frozen hands and toes.

This work would never have been completed without the love and support of my wife, Diane; parents, Lowell and Doris; and daughter, Laura, along with Sean, Jared, and the rest of my family.

Counties of Minnesota

Introduction

Minnesota's Northern Pike Legacy

The "aquatic wolf" was a creature of awe and mystery in the Old World, with early poems and folk tales alluding to its large size and predatory nature (Casselman 1996). Fearsome stories were told about northern pike attacking people and other creatures, and an Ostrogoth king reportedly saw the visage of his recently

FIGURE I.1. Northern pike displayed on the wall at historic Grand View Lodge, Nisswa, Minnesota.

executed enemy in a pike's face (Hoffmann 1987). Christian religious symbolism was even imputed to markings on the heads of pike. Although many of the early European myths and legends have been lost to us, pike still play a unique role in our cultural heritage. Large pike project an intriguing image of outdoor Minnesota and our landscape of more than 10,000 lakes. Huge fish are prominently displayed at sport and travel shows, and a visitor to nearly any bar or resort in rural Minnesota will find large pike mounted on the wall and unusual and unlikely stories told about how they got there (Figure I.1). Pike are such an icon of the north woods that beer commercials featuring pike fishermen enjoying a day on the Crow Wing chain of lakes even claim "It doesn't get any better than this!"

The more plentiful small northern pike have not held the same favored status in the minds of many anglers, who consider small pike a nuisance and blame them for interfering with the catch of other fish, such as walleye *Sander vitreus*. Therein lies one of the most challenging contrasts for managing the fish: large pike are prized catches whereas small pike are frequently regarded as trash fish. Unfortunately, fishing pressure has pushed many pike populations toward smaller-sized fish, and fisheries managers have struggled to reverse the trend. Regardless of size, pike are still highly valued as food. Well-prepared fillets of pike are described as sweet, white, and flaky.

As the most widely distributed game fish in Minnesota, northern pike provide recreational opportunities for summer and winter angling and the unique winter sport of darkhouse spearing. A diverse clientele uses the resource, including people interested in catching a meal of fish, casual anglers, and people intensely focused on catching trophy-size fish. After the walleye (Minnesota's state fish), pike are the most highly sought game fish by anglers, and they comprise a large proportion of the total weight of fish harvested in Minnesota each year. The incidental catch of pike is also an important bonus for many anglers not specifically fishing for them. When annual expenditures on recreational fishing in Minnesota (estimated at $2.5 billion in 2006) are considered, it is evident that the economic impact of pike in rural areas is considerable.

Northern pike populations represent a diverse resource within Minnesota. Because pike population dynamics reflect local geology and human influences, their populations and management vary considerably along a gradient from southwest to northeast within the state, and even from lake to lake. Pike populations have withstood subsistence, commercial, and recreational fishing since the beginning of recorded history. Nevertheless, pike fisheries and their management have changed considerably over the past 100 years in Minnesota, and recreational fishing during the past 60 years has caused noticeable changes in fish sizes and numbers that triggered efforts to actively manage or rehabilitate pike populations. Novel forms of fisheries management were developed to cope with the changes, and pike management has evolved in concert with advances in our understanding of the important dynamics of pike populations, the relationships between pike and their habitat, and the role of pike in fish communities. As important game fish, pike populations face increasing pressure as urbanization spreads and the human population in Minnesota continues to grow (Figure I.2). Pike populations are facing loss of critical habitat for spawning and nursery areas, overexploitation of large fish, and human degradations of their habitat such as pollution, invasion of exotic plants and animals, lakeshore development, and impoundments that affect water levels and flow. Recreational anglers request more stocking programs, special size regulations, and recovery of stunted populations and demand that the Minnesota Department of Natural Resources (MNDNR) fully evaluate regulations and management practices. Increasingly, management of this fish must be justified in legal and political arenas.

Much innovative research on northern pike management has been conducted in Minnesota. Pioneering work has focused on stocking success, fish community responses to stocking, management of rearing marshes, environmental effects on natural production, and pike genetics. Minnesota has also been a leader in developing recreational fishing regulations to improve the sizes of pike and testing experimental approaches to expanded bag limits, slot length limits, and maximum size limits. Unfortunately, some of the hard-won knowledge about pike management has already been lost. For example, important details have been lost about silver pike and northern pike hybrid culture and stocking programs that were conducted at a hatchery near Nevis, Minnesota, in the 1930s. Other remnants of our experiences with pike still exist in reports and documents scattered among a host of departmental records in Minnesota and in other states and provinces. In many cases, the important "kernels" of information are buried in the technical detail of these records. Preservation of such information is crucial to avoid repeating efforts and, more importantly, mistakes of the past.

The evolution of northern pike management strategies used by the MNDNR is a unique story of how our perceptions of the fish have changed through time. This book documents that story and describes the current state of our knowledge about the ecology, management, and sampling of pike in Minnesota, placing it within the context of research conducted elsewhere throughout the range of pike. The native range of pike includes northern portions of Europe and Asia, as well as North America, so this book synthesizes a large volume of technical ecological and management data. My purpose in this book is to provide a succinct synthesis of important and relevant historical literature and to incorporate my own work, which has focused on biology and management of northern pike during my 25-year career with the MNDNR.

FIGURE I.2. Recreational catch of northern pike from the 1950s.
PHOTOGRAPH COURTESY OF THE ITASCA COUNTY HISTORICAL SOCIETY PHOTO COLLECTION.

Chapter 1

The Ecology of Northern Pike

What's in a Name?

Northern pike entered the consciousness of Old World scholars as early as the 12th and 13th centuries. Scientific nomenclature for pike was assigned by the Swedish botanist Linnaeus in 1758 during his effort to design a classification scheme for all living organisms. In the New World, North American pike were originally thought to be distinct from those in the Old World, leading to different common and scientific names. Similarities between northern pike *Esox lucius* and muskellunge *Esox masquinongy* in North America added to confusion about the species during the 1800s (Scott and Crossman 1973). Though now accepted as a single species, different common names still exist for "northern pike" (in North America) and "pike" (in Europe) (Crossman and Casselman 1987). The scientific name is derived from a latinized Celtic term for salmon *(Esox)* and a common Roman personal name *(lucius)*.

The French adopted the common name *brochet*, literally translated as "poker" or "spear," which eventually became "pike" in English (Hoffmann 1987). Early legislation in Minnesota typically referred to the species as pickerel or great northern pike, and in everyday usages, small pike were commonly called pickerel. Other common names in North America have included great northern pickerel, northern, common northern pike, jack, jackfish, and snakefish. Among anglers, small, thin pike are often denigrated as hammer-handles, snakes, and even snot-rockets or slime (because of their extensive slimy coating).

Genetic Variation and Anomalies in Northern Pike

Relatively low genetic variability has been ascribed to the northern pike compared with other freshwater fish. Biochemical screening using allozyme and mitochondrial DNA genetic markers has detected few differences within and among populations over broad geographical ranges (Healy and Mulcahy 1980; Seeb et al. 1987; Senanan and Kapuscinski 2000). The reasons for such low levels of genetic variation are not clear but may include low effective population sizes (numbers of successful reproducing fish) and population bottlenecks associated with earlier glacial periods when the geographic range of pike was more limited (Miller and Senanan 2003; Craig 2008). The current lack of discernable genetic differences makes it difficult to develop genetics-based management strategies for pike. Recent use of microsatellite markers at the University of Minnesota has provided a more promising tool for examining genetic variation and differentiating among populations (Miller and Kapuscinski 1996; Senanan and Kapuscinski 2000; Miller et al. 2001). Microsatellites are numerous tandem repeats of short DNA sequences of 1–6 base protein pairs. Advantages of microsatellite markers are their relatively high level of variability and that small amounts of degraded tissue (such as found on pike scale samples) can be used as the source of DNA (Miller and Kapuscinski 1996). Using microsatellite techniques, it has been possible to differentiate among pike populations from different continents (Finland, Siberia, and North America), between brackish water and freshwater populations in Finland, and between populations in Alaska versus those in the north-central United States (Senanan and Kapuscinski 2000; Miller and Senanan 2003).

A color variant called silver pike occurs sporadically throughout the natural range of northern pike. Silver pike lack the markings of other northern pike, having a more solid silver or silver-blue background color (Figures 1.1-1.3). According to Eddy and Surber (1947), silver pike were first observed about 1930 in Lake Belle Taine near Nevis, Hubbard County. Local anglers reported that silver pike did not appear in their catches until about that year. Silver pike and muskellunge were propagated for several years in the Nevis hatchery and stocked into nearby lakes. Since then, silver pike have been noted from at least 18 lakes across northern Minnesota, and many of those lakes were never stocked with Nevis fish.

A study in which silver pike were raised up to 3 years old at the University of Minnesota (Eddy 1944) and breeding experiments at the Spirit Lake, Iowa, hatchery (Moen 1962) showed that mixing of silver pike eggs and sperm produced silver pike offspring. Mixing silver pike with regular northern pike produced young with poor survival that had unusual black mottled splotching on the body. A recent biochemical study using mitochondrial DNA (Foster 2000) was unable to distinguish between silver pike and northern pike. In view of those findings, it is likely that the silver pike is simply a genetic color variation.

FIGURE 1.1. Silver pike below a typical northern pike trap netted from Young Lake, St. Louis County, in May 1993.
FILE PHOTOGRAPH FROM MINNESOTA DEPARTMENT OF NATURAL RESOURCES.

FIGURE 1.2. Silver pike caught in the Cullen chain of lakes, Crow Wing County, January 2010. Shawn Langhorst described the freshly caught 32-inch fish as stunning bluish gray on the sides with a black back, silver fins—and not a single spot on it.
PHOTOGRAPH COURTESY OF SHAWN LANGHORST.

Apart from the unusual silver pike, background color of typical northern pike can be somewhat variable, ranging from green to brown, and in one reported case, yellow-green. Several instances of a bright silver foreground color have been documented from Disappointment, Parent, and Snowbank Lakes in Lake County (Figure 1.4). Light, horizontal, bean-shaped spots on the sides are typical, but in Minnesota, chainlike patterns have commonly been observed (Figures 1.5–1.6). A study in England found that markings of adult pike were like human fingerprints; they were specific to individual fish (Fickling 1982).

A rare and truly bizarre anomaly dubbed the "horned pike" has also been encountered in Minnesota. The horns seem to be extensions of paired bones from the upper surface of the duck-

FIGURE 1.3. Silver pike observed at the Spirit Lake Hatchery, Iowa.
PHOTOGRAPH COURTESY OF WILLIAM WACKERBARTH.

bill-shaped snout of northern pike (Figure 1.7). Crossman (2004) noted that specimens of horned pike were remarkably similar from Manitoba, Ontario, and North Dakota, provinces and states that all border Minnesota.

Hybrids of northern pike include natural crosses with the grass pickerel *Esox americanus* and muskellunge (Crossman and Buss 1965; Casselman et al. 1986). The only closely related species to pike in Minnesota, however, is the muskellunge. The northern pike and muskellunge hybrid cross, called tiger muskellunge, has commonly been propagated to augment recreational fisheries in Minnesota. Stocking of the hybrids is considered a muskellunge management tool.

FIGURE 1.4. Northern pike with bright silver foreground colors angled from Snowbank Lake, Lake County, September 1986.
FILE PHOTOGRAPH FROM MINNESOTA DEPARTMENT OF NATURAL RESOURCES.

FIGURE 1.5. Northern pike with chainlike color pattern between two normally colored fish from Chad Lake, St. Louis County, August 1979.
PHOTOGRAPH COURTESY OF BRIAN KONTIO.

FIGURE 1.6. An unusual color pattern shown above a typical northern pike. The fish was from Big Trout Lake, Crow Wing County, August 2009.
PHOTOGRAPH COURTESY OF DICK HANKS.

FIGURE 1.7. Horned pike caught in July 1989 by Minnesota Department of Natural Resources survey crew in Pleasant Lake, St. Louis County.

FILE PHOTOGRAPHS FROM MINNESOTA DEPARTMENT OF NATURAL RESOURCES.

Distribution

Northern pike have a holarctic distribution, being commonly found in northern latitudes (35° to 75°N) of Europe, Asia, and North America (Crossman 1996). They are abundant in states and provinces of north-central North America, and Minnesota is centrally located in the natural range of pike within North America. Pike are the most widespread game fish in Minnesota; they are present in all major drainage basins in the state. Connections between waters probably explain much of the natural distribution of pike since the last deglaciation. A study of 1,029 lakes in the northern boreal region of Sweden found pike present in 95% of the 693 Swedish lakes that had high connectivity with other waters, but they were also present in 121 lakes that had barriers to pike movement (Spens and Ball 2008). Pike colonization of the Swedish lakes was not limited by in-lake morphology or water chemistry. In Minnesota, pike were probably not native to some highland lakes (for example, lakes in elevated areas north of Grand Rapids, Itasca County), but extensive stocking and transplanting extended their distribution within the state. Pike are now present in 3,351 waters throughout the state (including border waters) (MNDNR lake survey data) covering about 0.9 million hectares (2.2 million acres) or about 95% of the accessible lake acreage in Minnesota.

Northern Pike Habitat

Water Temperature and Dissolved Oxygen

Northern pike can tolerate a broad range of environmental conditions but thrive best in moderately productive mesotrophic (having moderate amounts of nutrients) to eutrophic (rich in nutrients) freshwater lakes and rivers. Pike seem to thrive equally well in the crystal clear water of trout lakes and in the darkly stained (tea-colored) water of lakes in the Kawishiwi River system, Lake County. They are considered cool-water fish and grow even in Minnesota winters when water temperatures are less than 5°C (41°F). The optimum temperature for growth in summer is 19°C to 21°C (66°F to 70°F) for adults and a few degrees higher for young-of-the-year individuals (Casselman 1996). Pike are seldom found where water temperatures exceed 28°C (82°F) (Figure 1.8). In two southern Ohio reservoirs, summer habitat for pike was restricted by warm surface water temperatures and hypoxic (extremely low levels of dissolved oxygen) bottom waters (Headrick and Carline 1993). When surface temperatures exceeded 25°C (77°F) during two to three months in summer, pike were found in the coolest water available that had dissolved oxygen concentrations of at least 3 milligrams per liter. In Pool 14 of the Upper Mississippi River, pike moved into backwater areas during fall when water temperatures were falling from 10°C to 5°C (50°F to 41°F) (Sheehan et al. 1994). Pike overwintered in the backwater areas, then left in spring when water temperatures started rising. Casselman (1996) provided an

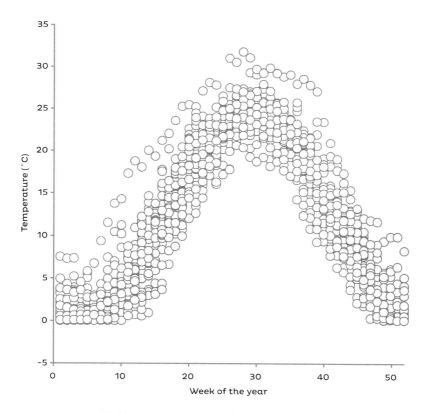

FIGURE 1.8. Weekly average temperatures from flowing (well-mixed) waters where northern pike occur.

DATA AND GRAPH ARE FROM A NATIONAL COMPENDIUM OF WATER TEMPERATURE AND FISH OCCURRENCE RECORDS PROVIDED BY J. EATON, U.S. ENVIRONMENTAL PROTECTION AGENCY, DULUTH ENVIRONMENTAL RESEARCH LABORATORY (EATON ET AL. 1995).

in-depth review of temperature preferences and requirements for juvenile and subadult pike.

An amazing technology has developed using transmitters that can be surgically implanted in fish; the transmitters emit information about the temperatures of the fish and depths of the water occupied by fish. In Little Wabana Lake, Itasca County, we implanted ultrasonic transmitters in northern pike and recorded signals from the transmitters at hydrophones stationed in the lake (R. Pierce, unpublished data). Signals from the ultrasonic transmitters are short sequences of sound, and the hydrophones act as listening devices that interpret the sequences of sound. The transmitters provided an enormous amount of information about internal body temperatures maintained by pike in a temperature-stratified lake. This was a unique opportunity to observe how pike choose thermal habitat because the lake offered well-oxygenated water as cool as 57°F (13.8°C) (at depths down to 8.5 meters [28 feet]) and surface temperatures that ranged up to 81°F (27°C) during the heat of the summer. Figure 1.9 illustrates the temperatures recorded from a small (484 millimeter [19.1 inch]) pike and a larger (810 millimeter [31.9 inch]) pike in Little Wabana Lake. The annual temperature series for the two pike in the stratified lake show remarkable similarity to temperatures recorded for pike in flowing, well-mixed waters (Figure 1.8).

Northern pike are tolerant of relatively low levels of dissolved oxygen, as evident by their overwinter persistence in some shallow lakes. Pike were able to maintain themselves in Lake Traverse, a shallow lake in Traverse County, even with winter oxygen levels that ranged as low as 0.9 to 2.7 milligrams per liter (Moyle and Clothier 1959). Juvenile pike can tolerate short-term (24-hour) exposures to dissolved oxygen concentrations as low as 1.2 milligrams per liter (Adelman 1969). Growth rate of juvenile pike can be decreased by reductions in dissolved oxygen concentration, but decreases in growth rates do not become extreme until concentrations of 3 to 4 milligrams per liter are reached. Adelman and

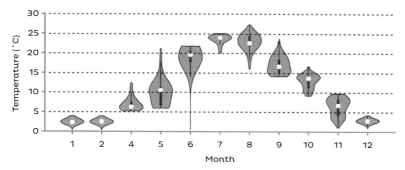

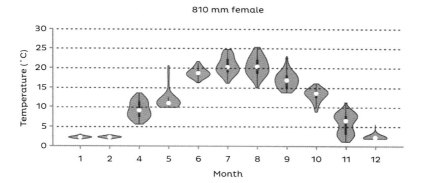

FIGURE 1.9. Monthly temperatures recorded from ultrasonic transmitters for a small (484 millimeters) and a large (810 millimeters) northern pike in Little Wabana Lake, Itasca County. The widest portions of violin plots are the most commonly observed temperatures; the narrowest indicate few observations at those temperatures; and open circles represent the middle of the temperature observations. No temperature readings were available for March.

Smith (1970) found that the decrease in growth was caused by reduced food consumption down to about 3 milligrams per liter, and by reduced food conversion efficiency below that. Oxygen tensions of less than 33% saturation are inadequate for survival of pike embryos and larvae (Siefert et al. 1973). Small pike have been considered to be more tolerant of oxygen depression than large individuals (Casselman 1996).

Visual observations in very low oxygen conditions during

winter found that northern pike were less active than fish such as yellow perch *Perca flavescens* and bluegill *Lepomis macrochirus*. In cage experiments, pike sought a position immediately below the ice, where dissolved oxygen concentrations were highest, and the fish remained in domes they formed by eroding the undersurface of the ice with fin movements (Magnuson and Karlen 1970). In our work with ultrasonic telemetry (R. Pierce, unpublished data), pike with transmitters generally remained in water having greater than 3 milligrams per liter dissolved oxygen concentrations. Nevertheless, an 807-millimeter (31.8-inch) female spent at least 47 minutes in water with less than 0.5 milligrams per liter dissolved oxygen (less than 4% saturation). That excursion went into water over 12 meters (39 feet) deep and occurred during mid-February in Little Wabana Lake.

Lake Basin Morphology

Northern pike populations are closely linked to the type of lake basin they inhabit. Considerable variety in pike population characteristics is found within Minnesota's geographical setting, which grades from prairie in the southwest, to a more mixed central area including hardwood forests, and then to heavily forested glacial shield in the northeast. Lakes along this southwest-to-northeast axis tend to grade from shallow, turbid, and eutrophic waters in the southwest to deeper, clearer, and oligotrophic (having low levels of nutrients) waters in the northeast, and fish communities shift accordingly along this geographical axis (Moyle 1956). Minnesota's thousands of lakes have been further categorized into 44 ecological lake classes based on variables associated with lake basin size, shape, and depth, variables associated with chemical fertility of the lakes, and length of the growing season (Schupp 1992). Within this wide range of ecological settings, it is difficult to generalize about pike population characteristics. However, populations from lakes in southern Minnesota tend to be more limited in natural recruitment (replacement through natural reproduction) due to loss of habitat from agricultural and other shoreland development. In contrast, many populations in the central and northern regions of the state have good natural recruitment that has historically supported a large recreational fishery.

In 16 small north-central Minnesota lakes, northern pike population sizes were more closely related to lake dimensions (morphometry) than to differences in other ecological factors such as water productivity (amounts of dissolved chemicals in the water and water clarity), recreational angling exploitation, or the relative abundance of prey fish (Pierce and Tomcko 2005). Density of pike (number of fish per hectare exceeding 350 millimeters [13.8 inches]) increased as the proportion of shallow littoral habitat in a lake increased (Figure 1.10). The littoral area of a lake is the shallow region where light reaches the bottom and aquatic plants grow, although in Minnesota, for convenience, we have historically defined littoral habitat as water less than 15 feet (4.6 meters) deep.

The relationship between northern pike density and the proportion of littoral habitat in a lake also fit for populations from other geographic areas (Wisconsin, Ontario, and Europe), except

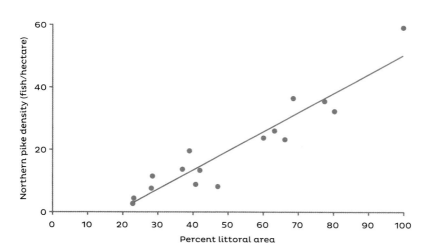

FIGURE 1.10. Relationship between percent littoral area of lakes and the density of northern pike populations from sixteen north-central Minnesota lakes.

when recruitment was possibly affected by conditions such as winterkill (dissolved oxygen levels that are too low to sustain fish over winter) (Pierce and Tomcko 2005). Because other factors such as winterkill can reduce pike density, the relationship illustrated in Figure 1.10 may predict the optimum density of pike that can be expected in a lake, and lower pike densities may be possible when other conditions limit density.

The amount of littoral area in a lake also seems to affect the sizes of northern pike in a lake. For example, a decline in the average size of individual pike was evident among Minnesota lakes that had a greater proportion of littoral habitat (Pierce and Tomcko 2005); shallow lakes with high pike densities typically contained smaller fish (Figure 1.11). A large-scale analysis of Ontario populations indicated that gill-net catch rates of pike during the fall were positively correlated with percentage of littoral area and negatively correlated with growing-degree-days above 5°C (Malette and Morgan 2005). Thus, pike catch rates in the Ontario lakes were highest in shallow water bodies in cool climates.

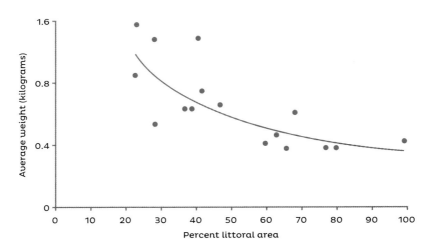

FIGURE 1.11. Relationship between percent littoral area and the average weight of individual northern pike sampled from sixteen north-central Minnesota lakes.

In contrast to shallow lakes with large amounts of littoral habitat, deeper and cooler lakes often have low densities but larger sizes of northern pike. A comparison of 14 northwestern Ontario lakes led Laine (1989) to conclude that pike populations found in deep, nutrient-poor oligotrophic lakes exhibited low population densities and superior growth rates, and were mainly limited by low amounts of available spawning and nursery habitat for juvenile pike. Populations from more nutrient-rich eutrophic lakes had greater densities, exhibited poorer growth rates, tended to be more opportunistic predators, and also appeared to be more limited by available food resources. In a study of fishing contest data from northwestern Minnesota, Jacobson (1993) found that production of trophy (greater than 6.8 kilograms [15 pounds]) pike was greatest in lakes with deep, cool water that also supported cisco *Coregonus artedii*. Pierce and Tomcko (2005) found that numbers of large (greater than 600 millimeters [23.6 inches]) pike in small lakes were more closely related to total surface area or the length of the shoreline than to littoral area. Total area or shoreline length by themselves explained 76% to 85% of the variation in numbers or total weight (biomass) of pike larger than 600 millimeters (23.6 inches) among lakes. A predictive model that included total area, shoreline length, and thermal habitat area (a measure of how much space was available to pike that also had optimum temperatures for growth) explained 97% of the variation in numbers of large pike among lakes (Pierce and Tomcko 2005). These findings parallel suggestions by Jacobson (1993) and Diana (1996) that pike exhibit shifts in types of habitat that are important as they grow to larger sizes. In all, the relationships between pike populations and the shapes of the lake basins they inhabit are important enough that basin dimensions offer some potential for predicting density, biomass, and even the sizes of fish in Minnesota pike populations (Pierce and Tomcko 2005).

Aquatic Plants

Throughout their lives, northern pike are associated with the aquatic vegetation that occurs in littoral and shoreline habitat (see the reviews by Bry 1996 and Grimm and Klinge 1996). A study of 640 Minnesota lakes found that lakes with high frequency of occurrences of diverse plant forms had the highest relative abundances of pike (Cross and McInerny 2006). Pike use a wide variety of plant forms as substrate for spawning, for juvenile nursery areas, and for hiding cover as adults. In a review of scientific literature, I found that a total of 39 plant genera were used by pike across Europe and North America. The plants most commonly referenced were sedges *Carex* spp. and grasses (including wild rice *Zizania*) for spawning habitat; milfoil *Myriophyllum* spp., pondweeds *Potamogeton* spp., and reed canary grass *Phalaris arundinacea* for juvenile habitat; and pondweeds for adult cover. Removal of near-shore aquatic plants is associated with destruction of nursery habitat for larval pike (Bryan and Scarnecchia 1992; Radomski and Goeman 2001; Radomski 2006) and occurs where shorelines are disturbed by construction of homes, boat docks, and swimming beaches. Inskip (1982) developed habitat suitability models for pike in which vegetation in the spawning habitat during spring, and submerged and emergent vegetation during summer were important components. Habitat suitability models condense observations about the capacity of the habitat to hold fish into a mathematical model driven by the most useful number of habitat measurements. The habitat suitability models for pike also incorporated water temperatures, pH, dissolved solids in the water, length of the growing season, water level fluctuations, and in rivers, the availability of slack water habitat.

The young northern pike seem to change the habitats they use as they grow in size. In a Danish lake, stocked young-of-the-year pike tended to use complex structure (created artificially with spruce trees) early in the summer but used less complex vegetated structure provided by reeds *Phragmites* and cattails *Typha* in late summer (Skov and Berg 1999). Underwater observations of pike in a small (23 hectares [57 acres]), but deep (40 meters [131 feet]) Alberta lake found that pike selected from several habitats available to them, relying most heavily on shallow vegetated areas in the lake (Chapman and Mackay 1984a) (Figure 1.12). However, pike larger than 250 millimeters (9.8 inches) were observed in deep, unvegetated areas more often than smaller, young-of-the-year pike, indicating that the older fish more fully used the whole range of habitats available in the lake compared to young-of-the-year. Radio and ultrasonic telemetry studies involve tracking of individual fish using transmitters that are implanted in the fish and emit radio or sound signals. Telemetry studies in a larger lake, Seibert Lake, Alberta (3,500 hectares [8,650 acres]), showed that pike moved extensively throughout the lake, but they exhibited a distinct preference for shallow vegetated areas close to shore (Chapman and Mackay 1984b). In Coeur d'Alene Lake, Idaho, large pike males that were radio-tracked used vegetated habitat (primarily pondweeds *Potamogeton* spp. and mixed stands of aquatic plants) 93% of the time even though less than 10% of the lake was vegetated (Rich 1992). Experimental gill-net catches of pike in two shallow Ontario lakes were greatest at intermediate plant densities (35% to 80% of submerged vegetative cover) (Casselman 1996; Casselman and Lewis 1996). The plants were primarily *Potamogeton* spp., coontail *Ceratophyllum demersum*, and Canada waterweed *Elodea canadensis*. Small pike were usually caught in the densest beds of these plants, and large fish were in more sparse vegetation.

Similarly, an inverse relationship was found between northern pike sizes and the density of aquatic plants in a small (27 hectares [67 acres]) Swedish lake, suggesting that the small pike stayed in dense vegetation to avoid interactions with larger pike (Eklov 1997). The largest pike in the Swedish lake were observed in or close to tree structure habitat. Small pike tend to avoid large pike, so the distribution pattern of pike in a lake is influenced by fish size and can be spatially clumped when pike densities are high (Nilsson

FIGURE 1.12. A northern pike blends into a bed of Robbins' pondweed *Potamogeton robbinsii*.
COPYRIGHT 2008 MINNESOTA DEPARTMENT OF NATURAL RESOURCES.

2006). In pooled sections of the Mississippi River below Minneapolis, submergent vegetation such as *Ceratophyllum demersum*, curled pondweed *Potamogeton crispus*, and *Elodea canadensis* was considered critical habitat for young-of-the-year pike (Holland and Huston 1984). With those important links to aquatic plants, it would seem that even small changes in plant cover could affect pike populations. However, removal of 6% to 11% of the aquatic plant cover in Mary and Ida Lakes, Wright County, was not enough to affect density, sizes, and growth rates of pike (Cross et al. 1992).

mainly in winter (Diana 1983a). During the cold-water season, northern pike ovaries develop large numbers of eggs, and fecundity (egg production) is directly proportional to the size of the female (Figure 1.15). A 1941 study by Vessel and Eddy estimated that individual females (ages 3 to 4) from Long Lake, Anoka County, and Mud Lake, Aitkin County, produced between 9,000 and 62,000 eggs. Females from Houghton Lake, Michigan, produced between 7,691 eggs for a 399-millimeter (15.7-inch) fish and 97,273 eggs for an 889-millimeter (35-inch) fish (Carbine 1943). Large females are capable of producing in excess of 200,000 eggs (Frost and Kipling 1967). Roughly 20,000 to 26,000 eggs are produced per kilogram (9,070 to 11,790 eggs per pound) of adult female body weight. In Lake George, Anoka County, Franklin and Smith (1963) developed the following regression of fish total length (in inches) and numbers of eggs produced:

number of eggs = 4,401.4 (fish length) - 66,245

where the total length of a fish is defined as the distance between the tip of the snout and the tip of the tail.

FIGURE 1.15. Egg-filled ovaries of a female northern pike just prior to spawning.

"Density dependence" is the phrase describing how the performance of animals (examples are fish growth rates and egg production) depends on the extent of crowding among individuals. Kipling and Frost (1969) provided some evidence that northern pike in Lake Windermere, England, had increased fecundity when their population numbers and biomass declined. Thus, density-dependent changes in egg production could be a potential population-regulating mechanism for pike. An extensive literature review of pike fecundity and egg development can be found in Raat (1988).

The peak of spawning movements for northern pike takes place in early spring, during and immediately after the time when ice melts from lakes. Spawning occurs from mid-March in southern Minnesota (44° N latitude) to early May in northern Minnesota (49° N latitude). The timing and duration of spawning activity are dependent on weather patterns; cold periods inhibit spawning whereas warming trends stimulate spawning. Pike move into shallow, warmer water as the ice recedes from shorelines. Immigration into a spawning area in Gilbert Lake, Wisconsin, occurred when water temperatures were 38°F to 51°F (3.4°C to 10.6°C) and coincided with ice breakup in the spawning area (Priegel and Krohn 1975). Spawning in Gilbert Lake was observed in afternoons when surface water temperatures were 50°F to 64°F (10°C to 17.8°C). In Lake George, Anoka County, most pike congregated in a spawning slough a few days prior to spawning when water temperatures exceeded 2°C (35.6 °F) (Franklin and Smith 1963). The spawning season in Lake George, observed over three successive years, lasted 4 to 19 days. Spawning at Lake George was observed in afternoons from 1400 hours to 1800 hours when surface water temperatures were 52°F to 63°F (11.1°C to 17.2°C). Water temperature, daily light intensity, and the presence of suitable vegetation as spawning substrate all operate together to stimulate spawning.

Northern pike commonly spawn in shallow bays, sloughs, and marshes where water temperatures warm earliest in the spring (Figures 1.16–1.17). Eggs are scattered, but highest egg densities are usually found in vegetation in water less than 2 feet (0.6 meters) deep (Figure 1.18). Pike eggs are sticky and adhere to vegetation in the spawning ground. In Lake Windermere, England, eggs were found on waterweed *Elodea*, stonewort *Nitella*, and milfoil *Myriophyllum* (Frost and Kipling 1967). The most heavily used shallow water spawning substrates in Minnesota and Wisconsin have included mats of sedges *Carex* spp., straw of wild rice *Zizania* spp., reed canary grass *Phalaris arundinacea* and other flooded grasses, spike rush *Eleocharis* spp., and wild celery *Vallisneria americana*. The highest egg densities in Lake George, Anoka County, typically occurred in dense mats of sterile culms (stems) of *Eleocharis* spp., and the next highest densities occurred along shore in stands of *Carex* spp. (Franklin and Smith 1963). In Bebe Pond, Wright County, eggs were most plentiful over dense tufts of *Carex lacustris*, whereas in Howard Pond, Wright County, eggs were found in a uniformly dense mat of reed canary grass *Phalaris arundinacea* around the perimeter of the pond (Adelman 1969). Greatest egg densities in six northern Nebraska lakes were found on flooded native prairie grasses, mowed hay, and even on broken hay bales (McCarraher and Thomas 1972).

Spawning in deeper water, to 3.6 meters (11.8 feet), has been reported in Lake Windermere, England (Frost and Kipling 1967); to 2.6 meters (8.5 feet) in Point Marguerite Marsh, New York (Farrell et al. 1996); and offshore in 2 to 5 meters (6.6 to 16.4 feet) in Rose Bay of the upper St. Lawrence River, New York (Farrell 2001). In Moose Lake, Itasca County, several miniature radio transmitters (implanted through the oviduct and into the egg masses of female pike just before spawning) were expelled on deep-water bars in 4.9 to 5.8 meters (16 to 19 feet) of water covered with muskgrass *Chara* and stonewort *Nitella* (Pierce et al. 2005). In complex habitats of the upper St. Lawrence River, New York, spawning begins in shallow seasonally flooded vegetation in tributaries and in shallow littoral habitat of bays (late March through early April), then shifts to deeper littoral areas of the bays and main river shoals by mid to

FIGURE 1.16. Spawning area of northern pike in Willow Lake, Itasca County.

late May (Farrell et al. 2006). Computer simulations of survival and growth of young-of-the-year pike indicated that spawning in the tributaries was much more productive than in shallow bay habitat, and very little production resulted from spawning in deeper littoral habitat (Farrell et al. 2006).

Recent evidence shows that northern pike can exhibit both spawning-site and natal-site fidelity. Tagging and genetic studies in Lake Kabetogama, a large 10,425-hectare (25,760-acre) lake in northern St. Louis and Koochiching Counties, have shown that individual fish return to the same spawning grounds where they hatched over consecutive reproductive seasons (Miller et al. 2001). Pike from two spawning sites, Daley Brook and Tom Cod Creek, located 14.8 kilometers (9.2 miles) apart, mixed during nonspawning periods in Lake Kabetogama but showed little straying when they returned to

FIGURE 1.17. Wetland complex for spawning northern pike in Ruby Lake, Itasca County.

FIGURE 1.18. Vegetated underwater substrate used by spawning northern pike in Ruby Lake, Itasca County.

natal spawning grounds. In contrast, pike appeared to move among several spawning sites in the much smaller 217-hectare (536-acre) Lake George, Anoka County, (Franklin and Smith 1963). In the upper St. Lawrence River, New York, Bosworth and Farrell (2006) found some evidence of genetic differences among spawning sites but could not confirm natal-site fidelity. Experiments along coastal waters of Finland and Sweden in the Baltic Sea relocated tagged pike to various distances away from their spawning areas. Homing toward their spawning areas was very pronounced when pike were relocated less than 10 kilometers (6 miles), but they did not orient back to their original spawning grounds if they were moved more than 60 kilometers (37 miles) away (Karas and Lehtonen 1993).

During the spawning act, individual females are closely attended by 1 to 3 males, which swim at her side (Clark 1950). Males and females roll toward each other so that eggs and milt can be released together, vibrating their bodies together to extrude and fertilize up to 60 eggs at a time (Becker 1983). The group then launches forward, scattering eggs across the substrate. Clark (1950) observed that spawning took place every 3 to 5 minutes for undisturbed groups. Clark also noted that one female, with a distinctive mark on her head, was observed spawning on 3 consecutive days.

Two northern pike diseases are often observed during the spawning period in Minnesota. The first is a herpesvirus (esocid herpesvirus-1) (Yamamoto et al. 1983) commonly referred to as "blue spot" because of the pale blue spots that temporarily form on the back, sides, and fins of pike during the spawning period (Figure 1.19). Blue spot is apparent in as much as a third of the population from individual lakes, but the blue spots are visible only during spring when water temperatures are 2°C to 13°C (36°F to 55°F); the

FIGURE 1.19. Esocid herpesvirus-1, as evidenced by pale blue spots on the side and dorsal (back) surface of a northern pike.

spots then typically disappear after spawning when water temperatures reach 14°C (57°F) (Margenau et al. 1995).

Lymphosarcoma is the second disease that is observed mostly in late winter or early spring during spawning. The disease is not completely understood but, bizarrely, has been considered a malignant tumor associated with a virus (Wyatt and Economon 1981). Lymphosarcoma tumors are large, firm swellings, welts, or lumps that resemble sores and have loose or missing scales over the crown of the tumor (Figure 1.20). The tumors have an inflamed red or slightly bloody appearance and are generally located on the sides of the fish behind the pelvic fins (rearmost set of paired fins). The disease seems to occur less frequently than herpesvirus but has been reported with an overall frequency of 21% of wild northern pike, which is the highest frequency of occurrence known for malignant cancer in any free-living vertebrate (Papas et al. 1976; Sonstegard and Hnath 1978). Lymphosarcoma tumors are often reported by ice fishers during the winter because tumors carried over into summer are hardly noticeable, regressing in size and changing in consistency as water temperatures warm. Warm water in the summer may inhibit the virus (Sonstegard and Hnath 1978). Lymphosarcoma is contagious, and transmission from fish to fish could occur at spawning time when infected cells are implanted into adjacent pike through body contact (Wyatt and Economon 1981), but the disease does not seem to be transmitted to the eggs (Sonstegard and Hnath 1978). Several test groups of pike yearlings that were injected beneath the skin with tumor homogenate (a solution having tumor cells) developed their own tumors in about one year, whereas control groups injected with a filtered, cell-free solution prepared from the same homogenate did not develop lymphosarcoma (Philip Economon, unpublished data, 1975, MNDNR Fish Pathology Laboratory, St. Paul, Minn.).

Eggs of northern pike are clear, amber colored, and 2.5 to 3.0 millimeters (0.10 to 0.12 inch) in diameter, and tend to blend into the vegetation where they are attached (Figure 1.21) (Franklin

FIGURE 1.20. Lymphosarcoma tumors on the side of a northern pike.
FILE PHOTOGRAPH FROM MINNESOTA DEPARTMENT OF NATURAL RESOURCES FISH PATHOLOGY LABORATORY.

and Smith 1963). Egg incubation takes about 2 weeks in Minnesota hatcheries but is highly dependent on water temperature. In natural spawning areas exposed to varying spring temperatures, eggs have hatched in 10 to 31 days, and estimates of survival to hatch range from 60% to 90% for the untended embryos (Bryan 1967; Adelman 1969). Wright and Shoesmith (1988) found a significant relationship between the length of female pike and the size of their eggs, but increases in egg size did not translate into larger sizes for the newly hatched fry in their study. Similarly, a more recent study found that egg diameter and the size of adult spawning females did not seem to be related to amounts of egg nutrients or egg and larval performance (Murray et al. 2008). However, measurements of egg weight were correlated with increased nutrients in the eggs and larger size of the swim-up fry that were produced. Murray et al. (2008) found that eggs of late-spawning pike weighed more and developed more rapidly to swim-up stage than eggs of early spawners.

Optimum water temperature for development of the egg stage, as determined from lab experiments, was 12°C (54°F), and temperatures should not drop below 7°C (45°F) or rise above 19°C (66°F) (Hokanson et al. 1973). Poor water quality where the eggs are laid, such as elevated levels of hydrogen sulfide, can be harmful to northern pike eggs and possibly sac fry (Adelman 1969). Johnson (1957) suggested that high spring water levels during spawning in Ball Club Lake, Itasca County, and a small decline in water levels during egg incubation represented favorable conditions for pike production. Estimates of year-class strength in two large Missouri River reservoirs (Lakes Oahe and Sharpe) indicated that large year classes resulted from stable to rising water level and temperature, availability of flooded vegetation, and calm weather during the spawning season (Hassler 1970; Nelson 1978). In contrast, smaller year classes in the reservoirs were associated with abrupt temperature changes, dropping water level, and silt deposits that covered the eggs. A habitat model attempting to explain the effects of water level regulation in the St. Lawrence River, Quebec, suggested that river discharges have had substantial influences on the amounts of suitable habitat for the eggs and on potential mortality following dewatering (Mingelbier et al. 2008).

Larval Development and Year-Class Strength

Northern pike are 6.5 to 8.0 millimeters (0.26 to 0.31 inches) long at hatching (Franklin and Smith 1960), and the larvae have numerous black-colored cells (melanophores) on their backs and on the yolk sac left over from the egg stage (Figure 1.22). An adhesive organ, located on front of the head, is useful for 4 to 10 days after hatching. The adhesive organ secretes mucous that attaches larval pike to aquatic vegetation or other material, suspending the larval fish

FIGURE 1.21. Northern pike eggs at (a) 48 hours and (b) 96 hours postfertilization at the Waterville hatchery.

PHOTOGRAPHS COURTESY OF DALE LOGSDON.

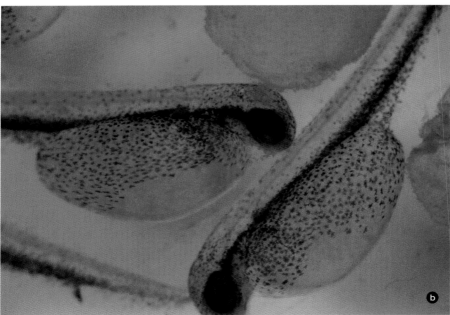

FIGURE 1.22. Northern pike fry (a) at time of hatching, (b) within one day of hatching, (c) two days posthatch, and (d) six days posthatch at the Waterville hatchery.
PHOTOGRAPHS COURTESY OF DALE LOGSDON.

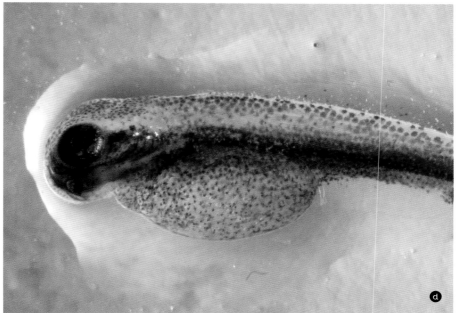

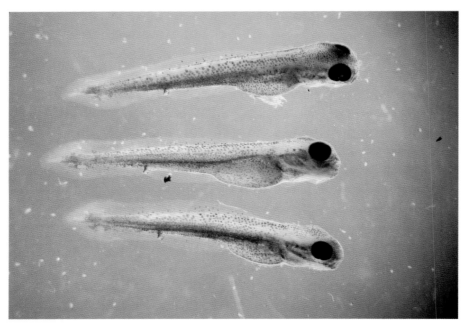

FIGURE 1.23. **Northern pike, 12-millimeter larvae.**
PHOTOGRAPH COURTESY OF STEVE SHROYER.

FIGURE 1.24. **Northern pike, 15-millimeter larvae.**
PHOTOGRAPH COURTESY OF STEVE SHROYER.

off the bottom and presumably keeping them out of the bottom silt and camouflaged from predators (Frost and Kipling 1967; Raat 1988). The early larval fish have long preanal (in front of the anus) finfolds with preanal measurements of 70% to 75% of total length; 8- to 13-millimeter (0.3- to 0.5-inch) larvae have myomere (body muscle segment) counts of 61 to 68 (41 to 46 preanal myomeres + 20 to 24 postanal myomeres) (Buynak and Mohr 1979). A pigmented band extends from the head to near the tail.

By 16 millimeters (0.6 inches), pectoral fin rays are visible, pelvic fin buds are present, and locations of the dorsal and anal fins are evident. Pectoral fins are the paired fins located immediately behind the head, and pelvic fins are the second set of paired fins midway down the body. The dorsal and anal fins are single fins originating on the back and underside of pike in front of their tails. In the Waterville Hatchery, Le Sueur County, personal observations show that the yolk sac is absorbed by about 13 millimeters (0.5 inches), and that larvae become active and start to feed at about 12

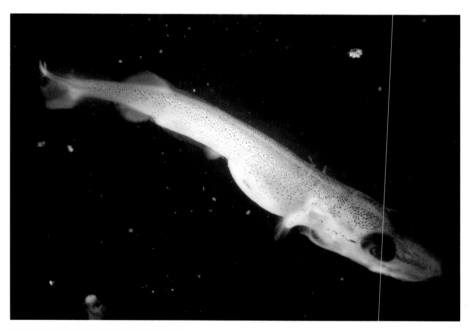

FIGURE 1.25. Northern pike, 20-millimeter larva.
PHOTOGRAPH COURTESY OF STEVE SHROYER.

millimeters. The characteristic duck-billed snout of northern pike becomes evident by 15 millimeters (0.6 inches). Fish larger than 20 millimeters (0.8 inches) are easily recognizable as esocid fry, and 30-millimeter (1.2-inch) fish have much of the characteristic shape of adults. Growth of the jaw and tail is more rapid than the body up to a length of about 65 millimeters (2.6 inches) because the jaw and tail are the most poorly developed regions of the fry (Franklin and Smith 1960). Pike larvae begin feeding 5 to 10 days after hatching, or about 4 days before the yolk sac is completely absorbed. Figures 1.23-1.25 illustrate development of larval pike from the Waterville hatchery.

Survival of the fry after hatching can be very low, and the time after hatching is a critical stage in year-class formation. Lab experiments found optimum temperatures for growth and survival that increased from 21°C (70°F) upon hatching to 26°C (79°F) for juvenile northern pike, but the upper level of temperatures they could tolerate was only 2°C higher (Hokanson et al. 1973). Water temperatures and chemical conditions were considered to be key factors for survival of young pike in a slough at Lake George, Anoka County (Franklin and Smith 1963). In Phalen Pond, Ramsey County, however, biotic influences such as available food and cannibalism were found to be more important than abiotic factors for fry survival (Bryan 1967). High mortality of fry in two rearing ponds in Wright County was also attributed to scarcity of food and cannibalism that occurred because of size differences among individuals (Adelman 1969). Survival after hatching was only 0.07% to 1.19% in the two ponds. The main effect of water quality was reduced growth rate due to low dissolved oxygen levels in one of the ponds. Danish studies further illustrated that size differences among fry can have large impacts on their survival during the first summer. Cannibalism among pike fry stocked into ponds in Denmark was greatest when the range of fry sizes stocked was greatest, and survival of stocked fry could be enhanced if the fry were all about the same size (Skov et al. 2003). Fry stocked early (10 May) grew larger and

FIGURE 1.26. Nursery habitat of northern pike fry and juveniles in Dove Bay, Rainy Lake, St. Louis and Koochiching Counties. Predominant aquatic plants in the bay were floating-leaf bur reed, wild rice, and a fringe of bulrush, sedges, and cattails. Wild rice and floating-leaf bur reed (close-up photograph) attracted the most pike fry.

had greater survival over the first summer (>12% survival) compared to smaller fry that were stocked 3 weeks later (<2% survival) in drainable ponds (Gronkjaer et al. 2004). The above examples show that the number of young pike produced does not necessarily depend on numbers of eggs spawned, and, therefore, year-class strength can be independent of the number of adults that spawn.

Movement of fry out of the spawning slough at Lake George occurred 16 to 24 days after hatching when the fry were about 20 millimeters (0.8 inches) long (Franklin and Smith 1963). Emigration out of the slough was influenced by light intensity, with bright sunny days triggering movement. Where spawning, nursery, and lake habitats are broadly connected (see Figure 1.26), northern pike movement is probably a more gradual shift in habitats than a distinct emigration.

Population Ecology

Sizes, Ages, and Growth Rates

The Minnesota hook-and-line record for northern pike is 20.75 kilograms (45 pounds 12 ounces; total length was not recorded) based on a catch reported from Basswood Lake, Lake County, in 1929, and Minnesota pike can exceed 1,150 millimeters (45 inches) in total length. The North American angling record is a 20.92-kilogram (46 pounds 2 ounces) fish caught in September 1940 in Sacandaga Reservoir, New York, and pike larger than 23 kilograms (50 pounds) have reportedly been caught in Europe. These behemoth sizes of record fish stand in stark contrast to the usual pike we encounter. The average total length of pike sampled during experimental gill-net surveys by the MNDNR has been 507 millimeters (20.0 inches), and weight about 0.8 kilograms (1.8 pounds) (Pierce et al. 1994). Fish exceeding 1,000 millimeters (39.4 inches) are seldom encountered in the nets. Creel surveys (studies where clerks at the lake monitor fishing effort and success) show that about 90% of angler-caught fish are between 380 millimeters (15 inches) and 635 millimeters (25 inches) long (0.3 to 1.6 kilograms [0.7 to 3.5 pounds]) (Cook and Younk 1998).

The first person who learned to determine ages of northern pike was Hans Hederström from Sweden (Hederström 1959; Jackson 2007). Hederström cleaned and dried pike vertebrae, finding concentric layers he attributed to annual growth. The discussion of his remarkable work was originally published in Swedish in 1759 and then republished in English in 1959. Nearly 250 years later, we have not been able to interpret ages much better than Hederström did. Sizes at age from his work were 450 millimeters (17.7 inches) at age 3, 600 millimeters (23.6 inches) at age 4, and 750 to 900 millimeters (29.5 to 35.4 inches) at age 6. He determined that 1,200-millimeter (47-inch) fish would be about 12 to 13 years old.

In Minnesota, northern pike ages have historically been determined from annual marks, or checks, on their scales. Scales first form in the midlateral region of the body along the lateral line when the young pike reach 32 to 35 millimeters (1.3 to 1.4 inches) (Casselman 1996). Scales from just above or below the lateral line at that midsection of the body are also some of the most uniform in size and shape (Williams 1955), and therefore scales from that area are typically used for age and growth studies. Pike scales are cycloid (have an evenly curved border), and in mature fish, the anterior portions of each scale (as taken from the sides of the body) have three distinctive lobes that overlap. Scales that regenerate after the fish's body has been damaged have been found in 7% to 16% of scale samples, and higher proportions of regenerated scales have been found in large fish compared to small fish (Williams 1955). In Michigan, the annual marks (annuli) are laid down in early March through May in the southernmost part of the state, and in late March through late June in more northern locations (Williams 1955). Identification of the first annulus is often difficult due to changing growth patterns during the first year of life that have been attributed to a shift in diet from invertebrates to fish. The 1955 dissertation by John Williams from the University of Michigan provided an intensive study of scale samples from known-age and partly known-age fish. Figures 1.27–1.29 are from a series of photos he sent to the Minnesota Conservation Department. Williams found that the first annulus often has a chainlike or bubbly appearance that resembles the inner portions of scales showing signs of regeneration.

Food availability and consumption rates influence the forma-

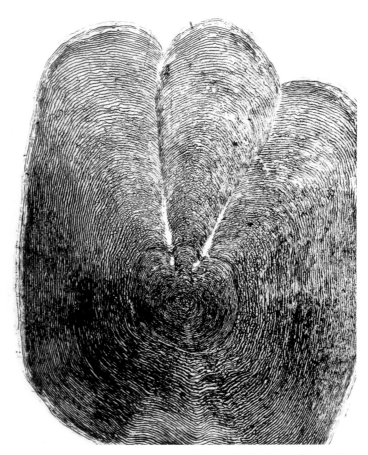

FIGURE 1.27. Scale from a known-age 351-millimeter (13.8-inch) yearling northern pike collected on May 3, 1951. Scales from the same fish on October 13, 1950, showed no check near the edge, so the check at the outer edge is the 1951 annulus. Note the appearance of a growth-rate change (near the center, or focus, of the scale) that occurred during July 1950.

PHOTOGRAPH COURTESY OF THE INSTITUTE FOR FISHERIES RESEARCH, MICHIGAN DEPARTMENT OF NATURAL RESOURCES AND UNIVERSITY OF MICHIGAN.

tion of annuli and growth of scales and bones. John Casselman's (1978a) comprehensive research on northern pike growth found that when adequate food resources are available, temperature influences food consumption through effects on appetite and metabolism. Casselman artificially induced false annuli in fish held in a laboratory by gradually increasing water temperatures after a period of greatly reduced growth.

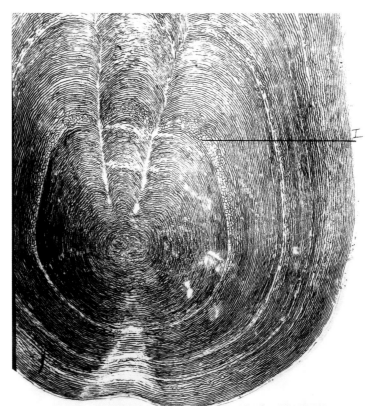

FIGURE 1.28. A typical first annulus (I) on a northern pike scale. Note the chainlike pattern of rapid scale growth after the first annulus was formed and its similarity to the pattern of regenerated portions of northern pike scales.
PHOTOGRAPH COURTESY OF THE INSTITUTE FOR FISHERIES RESEARCH, MICHIGAN DEPARTMENT OF NATURAL RESOURCES AND UNIVERSITY OF MICHIGAN.

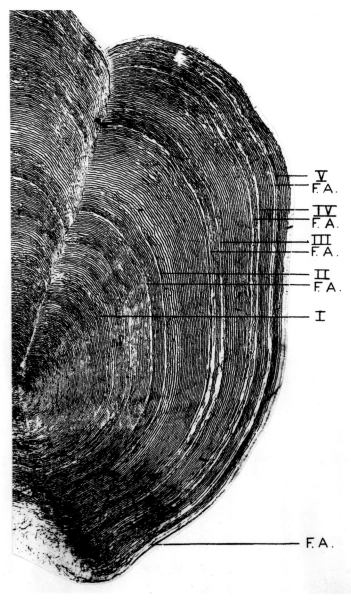

FIGURE 1.29. Scale from a 787-millimeter (31-inch) northern pike collected on August 20, 1953. Note the paired true (designated by Roman numerals) and false (F.A.) annuli that sometimes appear as "tracks." Tracking is most pronounced at ages II, III, and IV but is separated by only one circulus at age V. The 1953 false annulus, which would pair with the 1954 true annulus, has already formed at the outside edge of the scale.

PHOTOGRAPH COURTESY OF THE INSTITUTE FOR FISHERIES RESEARCH, MICHIGAN DEPARTMENT OF NATURAL RESOURCES AND UNIVERSITY OF MICHIGAN.

Unfortunately, the annuli and other checks on northern pike scales are notoriously difficult to interpret, especially from slow-growing fish, and older pike often exhibit fewer annuli on their scales than the true age of the fish. As a result, other calcified structures such as the cleithrum (a flat, L-shaped bone rearward of the gill cover; Figure 1.30), operculum (gill cover), and otolith (small stone-like structure within the inner ear) have been used extensively to help determine pike ages (Casselman 1990; Casselman 1996). The cleithrum is a bit easier to remove from the fish than the operculum, but a disadvantage of both bones is that the fish must be killed to remove them. The principal advantage of using bones is that the annual marks for older ages are easier to interpret than those on scales (Casselman 1990). Scale growth for old fish slows to the point that annuli on scales for older ages are usually obscure, whereas annuli for ages greater than 8 to 10 years are easier to observe in cleithra. The oldest age of a pike interpreted using modern techniques was that of a 29-year-old female angled in 1974 from Lake Athabasca, Saskatchewan (Casselman 1996). The fish weighed 14.2 kilograms (32 pounds) and had a fork length (measurement to the fork of the tail) of 1,100 millimeters

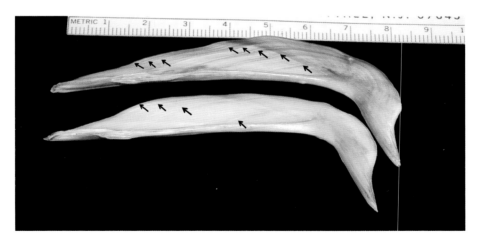

FIGURE 1.30. Cleithral bones from *(above)* a slow-growing male (674 millimeters [26.5 inches] total length; age 8) and *(below)* a similarly sized fast-growing female (670 millimeters [26.4 inches] total length; age 4) collected August 2000 in Lake Thirteen, Cass County. Arrows indicate where the annual translucent marks are located.

(43.3 inches). Minnesota populations consist primarily of fish less than 6 years old, and fish older than 12 do not seem to be common.

Growth of northern pike is generally fastest during their first year of life and progressively slows with successive ages. Sex of the fish also influences growth. Females may grow more quickly than males (Figure 1.30) and generally reach a larger ultimate size. Nevertheless, males as large as 1,110 millimeters (43.7 inches) long have been trap netted by MNDNR crews in the spring. Seasonally, growth of individual pike is most rapid when temperatures are increasing during spring and early summer, then decreases through late summer and fall, and is relatively low through winter (Casselman 1996). Growth rate can vary greatly among populations in Minnesota and elsewhere. Table 1.1 shows minimum, median, maximum, and interquartile ranges of total lengths at the beginning of each year of life. Those lengths were back-calculated from distances to each annulus measured on scale samples of each fish, with the proportional scale measurements applied to the length of the fish when it was captured. The pike scale samples were obtained from 298 Minnesota lakes (data from Jacobson 1993). The table also shows minimum, mean, and maximum total lengths at age from 82 lakes of worldwide circumpolar distribution (data from Casselman 1996). Pike from Europe and Asia attain about 13% greater weight for a given length than their North American counterparts (Doyon et al. 1988).

Differences in the growth rates of individual northern pike are influenced by fish density and environmental characteristics of the waters they inhabit. An analysis of growth rates for Minnesota populations showed that small, young pike grow fastest in shallow, fertile lakes with long growing seasons (Jacobson 1993). In small northern Wisconsin lakes, growth rates were negatively related to pike density, productivity of the water (as indicated by water clarity), and high abundances of small bluegills (Margenau et al. 1998). Pike growth in the Wisconsin lakes was also limited by extensive littoral habitat, high water temperatures (>21°C [70°F]) in summer, and winterkill. Growth and mortality rates in a northern

TABLE 1.1.
Back-calculated total lengths (millimeters) at scale annulus formation, sexes combined, for northern pike.

	Age									
	1	2	3	4	5	6	7	8	9	10
Minnesota populations										
Minimum	53	76	106	233	313	375	418	439	564	722
25%	191	325	406	463	506	548	602	673	773	866
Median	228	375	461	521	578	642	717	776	837	891
75%	254	430	515	583	646	721	782	839	876	919
Maximum	576	706	795	883	913	964	1,006	1,028	1,071	1,106
Circumpolar populations										
Minimum	99	165	239	304	348	402	452	466	496	581
Mean	192	332	437	516	584	649	667	715	745	795
Maximum	356	535	668	775	848	902	932	979	1,021	1,150

Ontario river were similar to those of a lake population at the same latitude (Griffiths et al. 2004), suggesting that current in the river seemed to have little effect on growth and mortality compared to other factors.

Population density, in particular, has a large effect on growth rates when individuals are forced to interact among themselves or compete for a limited food supply. In small Minnesota and Wisconsin lakes, growth rates are typically slowest when northern pike population densities exceed 12 to 14 fish per hectare (5 to 6 fish per acre) of fish sizes larger than 350 millimeters (13.8 inches) (Pierce, Tomcko, and Margenau 2003). Growth rates seem to increase sharply at lower densities. Figure 1.31 illustrates the nonlinear form of the relationship between growth and pike density.

The top panel of Figure 1.31 shows an example of the relationship between density and a simple index of northern pike growth, which is the average (mean) back-calculated length of the fish at each age (the example uses back-calculated length for age 4 males). The bottom panel illustrates an example of the relation-

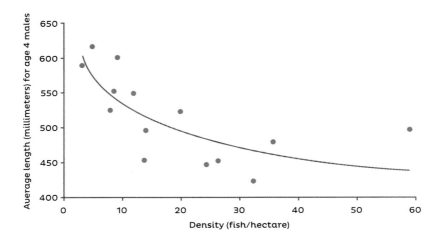

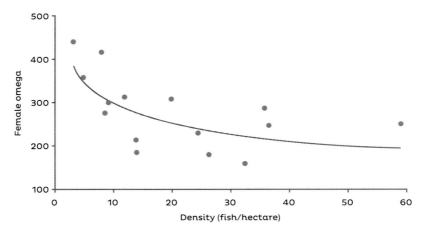

FIGURE 1.31. Average back-calculated length of age 4 males *(above)* and omega for females *(below)* as examples of northern pike growth rates in relation to pike density (number of fish longer than 350 millimeters per hectare) for fifteen north-central Minnesota lakes. Omega is the product of two parameters ($L_{infinity}$ and K) estimated from a mathematical model called the von Bertalanffy growth model. Details of the growth model are in Ricker (1975).

ship between density and a more complex measure of growth, omega, which is the product of parameters derived from the von Bertalanffy growth model (Ricker 1975; Gallucci and Quinn 1975) (the example uses growth parameters for female pike). The curvilinear shape of these relationships is largely an artifact of the lake area used to calculate density. When littoral area (rather than the total surface area) was used to calculate density, the negative relationships between growth rates and density were much more linear. These shape differences reflect the underlying ecological interrelationships between pike populations and the morphology of the lake basins they inhabit, and indicate that density should be standardized by the littoral habitat actually occupied by the fish stock. Regardless of how density is calculated, the important point is that differing densities explained a large amount of the differences in pike growth rates among lakes (Pierce, Tomcko, and Margenau 2003).

Grimm and Klinge (1996) also described a negative relationship between density and growth for young-of-the-year northern pike stocked into seven 0.1-hectare (0.2-acre) drainable ponds in the Netherlands. The fish were stocked into the ponds at different densities in mid-May, and their lengths and numbers were recorded when the ponds were harvested in early October. An outstanding example of how competition for limited food resources limits pike growth was observed in a field experiment in two small northwestern Wisconsin lakes. For the experiment, Margenau (1995) transferred pike from a high-density population (30 fish per hectare [12 fish per acre]) in Island Lake to a low-density population (4 fish per hectare [1.6 fish per acre]) in Largon Lake. Transferred fish demonstrated greatly improved growth in 1 year. Slow growth and condition of pike in Island Lake were considered symptoms of inappropriate food resources. Even when density dropped to 13 fish per hectare (5.3 fish per acre) in Island Lake, growth did not improve. Individual pike can exhibit high diversity in growth rates even within a population. Fry stocked into a Michigan rearing pond grew to a wide range of sizes, presumably because some individuals got an early competitive advantage in feeding (Carbine 1944). Subsequent growth of small pike in rearing ponds also furnished an example of growth compensation; the smallest fish at the end of the first year became the fastest growing individuals in succeeding years (Carbine 1944).

Growth rates of northern pike in natural environments are an integration of the amount and types of food consumed over time. Bioenergetics models are mathematical tools for relating what has been consumed to growth rates using information about physiological requirements of the fish along with temperature and growth data from the field. Diana (1987) used a bioenergetics model to investigate factors that could lead to stunting (slow growth and small sizes) of pike. Simulations with the model indicated that reduced amounts of available food, a lack of appropriately sized prey, and excessively warm water temperatures in summer could all cause stunting. A bioenergetics model was used by Headrick (1985) to study growth of pike in southern Ohio impoundments where summer water temperatures became very warm for pike. Modeling results suggested that the combined effects of warm water temperatures and the rates at which large pike could consume prey items limited their growth. In the warm water experienced in southern Ohio impoundments, large pike may not have been able to consume enough food to reach their growth potential and could even lose weight during summer. By moving to areas of the impoundments that had the lowest available water temperatures in midsummer, large pike had lower body maintenance requirements and were able to consume enough food to possibly channel more energy into growth. In contrast, small pike at the same warm temperatures had different consumption requirements and may have been able to consume enough food to exceed their maintenance requirements and continue to grow.

Density dependence in growth rates is often considered to result from a lack of food resources for the fish, but an intriguing study from Europe (Edeline et al. 2010) has provided new evidence that social stresses, simply from the way individual pike interact with each other, can be just as important. Edeline et al. (2010) stocked pike at different densities into experimental ponds but manipulated the amount of food the pike were receiving so that feeding rates were similar among the fish. Social stress among pike stocked at different densities into the experimental ponds caused changes in blood chemistry and a 23% reduction in body growth rate despite a similar food intake among the groups of pike. Edeline et al. (2010) extended their results to predict how social stress might impact pike population dynamics in Lake Windermere, England. They concluded that social stress can be a primary driver of pike population dynamics by affecting growth rates and reproductive potential.

Northern Pike Densities

Northern pike population densities are critically important in population dynamics and for management of the species. High levels of natural reproduction can cause high-density, slow-growing pike populations that are infamous among fisheries managers because of their inability to provide fish large enough to interest recreational anglers. Such populations seem capable of producing only "hammer-handle" size fish. Although high-density populations are common in small lakes of central and northern Minnesota, a wide range of pike densities can be found among natural populations of northern pike. A sample of 16 north-central Minnesota lakes (with surface areas of 15 to 765 hectares [37 to 1,890 acres]) had population densities that ranged from 3.2 to 59.0 fish per hectare (1.3 to 23.9 fish per acre) and biomasses of 3.6 to 33.6 kilograms per hectare (3.2 to 30.0 pounds per acre) for fish exceeding 350 millimeters (13.8 inches) total length (Pierce and Tomcko 2005). Mean density among the 16 populations was 20.7 fish per hectare (8.4 fish per acre), and mean biomass was 14.6 kilograms per hectare (13.0 pounds per acre).

Two fish removal efforts illustrated that northern pike can form very dense populations, especially considering they are a top-level predator. An unusual pike removal effort featured a 0.77-hectare (1.9-acre) gravel pit lake in Koochiching County. A total of 72 pike were removed during spring 2005, which implied a minimum density of 93.5 fish per hectare (37.8 fish per acre). The second example featured intensive netting of pike in Seth Lake, a 49-hectare (121-acre) Aitkin County lake, where 54 fish per hectare (21.9 fish per acre) or 27.6 kilograms per hectare (24.6 pounds per acre) were removed during spring and fall 1992. Additional netting during spring and fall 1993 yielded another 24 fish per hectare (9.7 fish per acre) or 12.8 kilograms per hectare (11.4 pounds per acre) for totals of 78 fish per hectare (31.6 fish per acre) and 40.4 kilograms per hectare (36.0 pounds per acre). Gill-net surveys showed that the pike population in Seth Lake was temporarily reduced through the removals but recovered within a few years owing to good natural recruitment. Gill-net catch rates in Seth Lake averaged 13.7 pike per net in 1991, 5.7 pike per net in 1994, and 12.4 pike per net in 1996.

The largest sizes of northern pike are much less abundant. In the 16 lakes described by Pierce and Tomcko (2005), progressive and pronounced decreases in density and biomass were found for larger-size fish. Mean density was 5.9 fish per hectare (2.4 fish per acre) for fish larger than 500 millimeters (19.7 inches), and was only 1.6 fish per hectare (0.6 fish per acre) for fish larger than 600 millimeters (23.6 inches) (Figure 1.32).

The relationship between density and the size structure (proportions of large fish) of northern pike populations is nonlinear (Figure 1.33) and is also similar in form to the relationship between density and growth rates (Pierce, Tomcko, and Margenau 2003). Size structure improves most at densities of less than 12 to 14 fish per hectare (5 to 6 fish per acre). The nonlinear form of the relationship between density and size structure may also be due, in part, to the underlying ecological interrelationships between pike populations and lake basin morphology. A more linear relationship was observed when density was calculated using littoral area rather than total surface area of each lake.

Because population size structure, growth rates, and secondary production rates are closely tied to pike population density, density seems to be the key link between the suitability of the habitat and

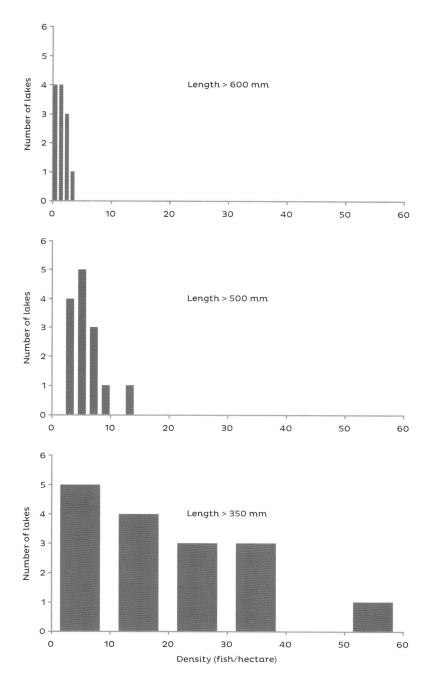

FIGURE 1.32. Densities of northern pike larger than 350 millimeters, larger than 500 millimeters, and larger than 600 millimeters total length among north-central Minnesota lakes.

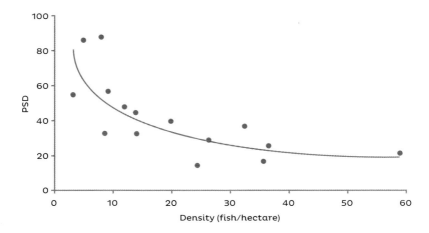

FIGURE 1.33. Northern pike population size structure plotted against density for twelve north-central Minnesota lakes. Proportional size distribution (PSD) is the percentage of fish larger than 350 millimeters that are also larger than 530 millimeters. PSD increases as the percentage of large fish in a population increases (Anderson and Gutreuter 1983).

the population dynamics of pike. The shape of a lake basin has strong influences on pike density, which in turn has important effects on competition, growth rates, secondary production rates, and the size structure of pike populations (Pierce and Tomcko 2005).

Mortality

A large proportion of the northern pike population dies each year, a rate that often exceeds 50%. Total annual mortality rates estimated for 17 northern Wisconsin lakes ranged from 35% to 79% (average = 57%) (Margenau et al. 1998). Total mortality includes all of the causes of death: both natural causes and fishing mortality. Other estimates from Minnesota and Wisconsin range from 45% to 91% (data from Johnson 1957; Snow and Beard 1972; Kempinger and Carline 1978; Snow 1978; Schramm 1983; Scheirer 1988; Goeman and Spencer 1992). In my own work, annual mortality estimated for 16 north-central Minnesota lakes ranged from 30% to 90%, and much of that mortality was likely to have been from natural causes because exploitation rates (fishing mortalities) from recreational fisheries were relatively low. A cautionary note about these total mortality estimates is that they can be biased if the ages determined for older pike are incorrect. Interpreting ages that are too young (based on the scales or bones of pike) leads to estimated mortality rates that are too high.

Tagging studies from 7 of the lakes demonstrated that recreational fishing caught only 4% to 22% of the pike that had been tagged (Pierce et al. 1995). However, recreational fishing for pike is very size selective, so the importance of exploitation increases substantially for large pike. Exploitation rates were two to nine times greater for large pike than for small ones. In one of the lakes, annual exploitation (the percentage of fish harvested each year) for fish larger than 500 millimeters (19.7 inches) was 46%.

Annual yields of northern pike to recreational fisheries in Minnesota have averaged 4.4 fish per hectare (1.8 fish per acre) and 5.0 kilograms per hectare (4.5 pounds per acre) over recent decades, and much of the harvest consists of fish ages 2 to 4 (Cook and Younk 1998; Pierce and Cook 2000). Yields from the earliest creel surveys in Minnesota were similar. A half century ago, from 1951 to 1956, annual pike yields averaged 4.8 kilograms per hectare (4.3 pounds per acre) among 32 lakes with intensive creel survey information (Johnson et al. 1957).

Relatively high total mortality compared with exploitation means that natural mortality is an important aspect of northern pike population dynamics, especially for small sizes of fish. This is significant because high rates of natural mortality have a large bearing on the effectiveness of different regulation strategies for managing pike. In some cases, cannibalism may be an important source of natural mortality for small pike. Cannibalism accounted for 30% of the large prey items in pike stomachs sampled from five northern Wisconsin lakes (Margenau et al. 1998). In a long-term study (1944-81) in Lake Windermere, England, cannibalism was most important when other food resources were scarce (Kipling

1983). In the River Frome, England, cannibalism accounted for most of the mortality of pike between 6 months and 2 years old (Mann 1982). Other European authors working in small water bodies (Grimm 1983; Treasurer et al. 1992) have suggested that cannibalism can be a mechanism for self-regulation in pike populations.

In Lake George, Anoka County, the highest natural mortality rate observed over several years occurred when the population density was also highest (Groebner 1964). After a 15-year study at Murphy Flowage, Wisconsin, Snow (1978) concluded that mortality was not density dependent except when pike density was made artificially high by stocking. From a review of available scientific literature, Allen et al. (1998) suggested that natural mortality and exploitation might be compensatory for small pike, so that fishing may substitute for natural causes of death in small pike. In contrast, the two forms of mortality may be additive for large pike so that fishing contributes additional mortality above and beyond the natural deaths of large pike.

Production Rates

Production rates have been regarded as important indicators of performance for fish populations, indicators that provide some suggestion of the amount of harvest that is sustainable. Production is a measure of the amount of fish tissue produced over time for a population and incorporates estimates of biomass, growth, and mortality. Unfortunately, production estimates for wild populations of northern pike are difficult to obtain and are therefore rare in the literature. Pike production estimates from Canada and Europe (expressed on a per year, or annual, basis) have included Squeers Lake, Ontario (production = 0.27 kilograms per hectare for ages 3 to 9) (Laine 1989); Lake Windermere, England (an average production rate of 1.5 kilograms per hectare over 19 years; fish ages 2 and older) (Kipling and Frost 1970); Lake Alinen Mustajarvi in southern Finland (2.6 kilograms per hectare; ages unknown) (Rask and Arvola 1985); Savanne Lake, Ontario (2.76 kilograms per hectare for ages 4 to 12) (Mosindy et al. 1987); and Lake Demments in western Russia (4.9 kilograms per hectare for ages 2 to 6) (Rudenko 1971). Annual pike production in two English gravel-pit lakes was 7.6 and 20.2 kilograms per hectare (ages 1 to 13) (Wright 1990), but the highest rate of pike production was reported from the River Frome in England, where annual production averaged 27.4 kilograms per hectare over 5 years (fish ages 2 and older) (Mann 1980).

In Minnesota, production estimates for northern pike were obtained from seven lakes in Aitkin and Itasca Counties. The annual production estimates ranged from 0.8 to 8.3 kilograms per hectare (0.7 to 7.4 pounds per acre of fish ages 2 and older), and averaged 4.1 kilograms per hectare (3.7 pounds per acre) (Pierce and Tomcko 2003a). A negative relationship was found between growth rate and production among the lakes (Figure 1.34), which is counterintuitive considering that production is calculated as the product of average biomass and instantaneous growth rates. An explanation can be found in density-dependent growth. Density-dependent growth was reflected in production estimates where the negative relationship between growth and production was more than offset by the strong positive relationship between pike density and pike production rate. Thus, the negative relationship in Figure 1.34 points to the overwhelming influence that density can have on production dynamics of pike, particularly for small pike.

Although the Minnesota data were limited to seven lakes, the resultant dome-shaped form of the relationship between northern pike density and production rate (Figure 1.35) appears to illustrate that production can be moderated by slower growth at high densities (Pierce and Tomcko 2003a). However, the simple curve illustrated in Figure 1.35 is heavily influenced by the more numerous young age groups (ages 2 to 4), and production dynamics can be more complex than the simple curve. For example, removal of 43% of the pike ages 2 and older from the population in Camerton Lake, Itasca County, during summer 1998 led to changes in growth rates, a rippling effect on recruitment and pike density, and an eventual

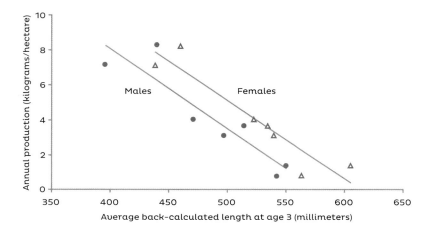

FIGURE 1.34. Individual growth rates for male (solid circles) and female (open triangles) northern pike, as measured by average back-calculated lengths at age 3 and annual production rates of pike in seven north-central Minnesota lakes.

shift toward larger sizes and older age fish in the population over an 8-year period. Although production of the youngest age groups in Camerton Lake was influenced most by population density (as in Figure 1.35), production of the oldest age groups (age 5 and older) seemed to be driven more by reduced growth rates. For example, production of age 5 and older pike in Camerton Lake was lower in 2006 (7.1 kilograms per year [15.7 pounds per year]) than in 1998 (10.3 kilograms per year [22.7 pounds per year]) even though their estimated population size was 2.5 times greater in 2006. These data support the commonsense notion that the best trophy pike populations will be found where environmental conditions support good growth of large pike. The Camerton Lake data also illustrated the complexity of production dynamics as pike density, growth trajectory, and their life span all shifted within a single population.

Our estimates of production rates seem to be low compared with annual yields (the weight of fish removed each year) for recreational fishing, and imply that recreational fisheries are harvesting much of the annual production of northern pike. The

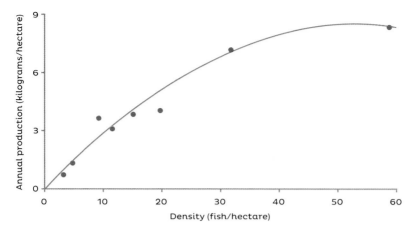

FIGURE 1.35. Density versus production estimates of northern pike age 2 and older from seven north-central Minnesota lakes.

average annual production estimate of 4.1 kilograms per hectare (3.7 pounds per acre) from the seven Minnesota lakes (Pierce and Tomcko 2003a) was lower than a statewide average annual yield for pike of 5.0 kilograms per hectare (4.5 pounds per acre) from the recreational fishery (approximated from 1980-96 creel surveys) (Pierce and Cook 2000). Although the average production rate and the statewide yield figures are merely averages, many pike populations in Minnesota maintain high densities of small individuals in the face of these yields. Pike production seems to be resilient to exploitation where spawning and nursery habitat are not limiting the numbers of young pike produced.

The cost of recreational fishing may come primarily from reduced production by larger, older fish. Annual production rates were found to be very low for older fish (Pierce and Tomcko 2003a). In exploited populations from the seven Minnesota lakes, most of the production occurred among the youngest ages and smallest sizes of northern pike (Figure 1.36). Very little of the total annual production occurred among the older, larger fish; the average

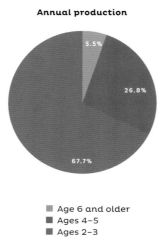

FIGURE 1.36. Proportions of the average total annual production estimated by age for ages 2 and older northern pike among seven north-central Minnesota lakes.

production rate for pike ages 6 and older was only 0.13 kilograms per hectare (0.1 pounds per acre). This is a very low number and demonstrates that larger, older pike can be very susceptible to overfishing. For perspective, in English units, it means that harvest of only one 10-pound pike from a 100-acre lake would remove the equivalent of an entire year's production of large pike from the lake. In this hypothetical example, removal of more than one trophy fish would deplete several years' worth of production. Since recreational fishing selectively targets large fish, which are the least productive part of the population, any management that is specifically aimed at producing large fish must severely restrict the harvest of large pike.

Trophic Relationships
Food of the Northern Pike

Since medieval times, northern pike have been considered "wolves among fishes" (Hoffmann 1987). They are lone, visual predators that ambush their prey from plants, stumps, and fallen logs. The long, slender body is well adapted to accelerating quickly and darting out after fish or other animals (Webb 1984). Fry feed first on microcrustacean zooplankton, but they rapidly progress to larger sizes of food. Early larval stages target copepods and water fleas (cladocerans), but by the time they reach 20 millimeters (0.8 inch) total length, they shift toward aquatic insects such as mayflies (ephemeropterans), damselflies (zygopterans), and midges (chironomids) and other forms of biting flies and mosquitoes (dipterans) (Hunt and Carbine 1951; Franklin and Smith 1963; Bryan 1967; Adelman 1969; Morrow et al. 1997). By the time pike reach 50 to 60 millimeters (2.0 to 2.4 inches), fish usually comprise most of their diet. In a Michigan spawning area, increased daily growth rate in June coincided with the appearance of larger food items, especially fish, in the diet of juvenile pike (Hunt and Carbine 1951). Variations in the diets of pike stem from changes in the relative influences of the types of prey available, food preferences, and the seasonal abundances of prey (Mann 1982). A long-term (1950–62) study of pike stomach contents at Heming Lake, Manitoba, found some seasonal variety in their diet and interannual differences in proportions of fish that had empty stomachs (Lawler 1965). However, the food habits of pike in Heming Lake did not change dramatically over an 11-year period of intensive fishing.

As the top-level predator in rivers and lakes, large northern pike are opportunistic and eat nearly any food item they can swallow. A resident of Ten Mile Lake, Cass County, reported catching a 2-pound pike that had even swallowed a large chromium-plated shower curtain ring (Carlson 2007). Fish are often more than 90% of the diet and include yellow perch, centrarchid sunfish, white sucker *Catostomus commersonii*, walleye, common carp *Cyprinus carpio*, numerous minnow species, trouts, burbot *Lota lota*, brook stickleback *Culaea inconstans*, bullheads *Ameiurus* spp., and other pike. Where pike and white sucker coexist, sucker abundance can be lower than in lakes without pike (Hinch et al. 1991). Coregonids, such as whitefish *Coregonus clupeaformis* and cisco, can be important diet items for trophy-size pike (Makowecki 1973; Jacobson 1993). Video recordings of tiger muskellunge (the pike × muskellunge hybrid) attacking fathead minnows *Pimephales promelas* illustrate how the fish coil into an S-shaped body posture to generate a fast, lunging strike (Webb 1984). The S-shaped posture allows a pike to attack at several angles (Webb and Skadsen 1980). Minnows are usually attacked from the side with the strike directed at the center of the prey fish's body, a point that moves least during escape attempts. Prey items are held with such strength by the jaws that mechanical jaw spreaders are needed to pry the jaws open when removing fishing baits. Prey fish are often swallowed head first but may also be swallowed tail first or sideways (Nursall 1973).

Invertebrates, such as insects (Chapman et al. 1989), leeches (Venturelli and Tonn 2006), and crayfish, are included in the diet and become more important when prey fish are scarce (Chapman and Mackay 1990). Comparative studies in northern Alberta lakes have suggested that individual northern pike within a population may even specialize on invertebrates in lakes where other prey fish species are available, although their growth may be stunted (Beaudoin et al. 1999; Venturelli and Tonn 2006). The crayfish *Orconectes virilis*, a fairly large food item, was second only to yellow perch in the diet of pike in Camerton Lake, a shallow bog lake in Itasca County (Pierce, Tomcko, and Drake 2003). Because crayfish were eaten during June and July, when pike were growing quickly, crayfish may have been making a significant contribution to the annual production of pike in Camerton Lake. Other vertebrates that happen to be available, such as frogs, rodents, and small birds, are occasionally victims of pike. Concerns about pike predation on young ducks in waterfowl production areas led to experiments showing that pike were not a consistent menace to the young birds, but the concerns were at least partly warranted based on studies of pike food habits (Lagler 1956). Daily food rations for pike are highest from May through August and are very low in winter. The rare inclusion of large food items in their diet has pronounced effects on the amount of calories pike consume (Diana 1979).

Yellow perch are often a key component of northern pike diets in natural systems of North America (Lawler 1965; Diana 1979). In Maple Lake, Douglas County, and Grove Lake, Pope County, pike fed mostly on yellow perch even when minnows and small sunfish were common (Seaburg and Moyle 1964). In lakes within Voyageurs

National Park, St. Louis County, yellow perch comprised more than 60% of pike diets in all seasons, and pike diets overlapped significantly with largemouth bass *Micropterus salmoides* diets in the park lakes (Soupir et al. 2000). Much of the growth and production of pike in Camerton Lake was attributed to yellow perch in the diet (Pierce, Tomcko, and Drake 2003), and availability of yellow perch was likely the most important factor affecting growth rate and body condition of pike in Escanaba Lake, Wisconsin (Inskip and Magnuson 1986).

Fish Community Interactions

Some researchers have considered the predator-prey relationship between northern pike and yellow perch to be an important controlling aspect of aquatic community structure in small lakes (Anderson and Schupp 1986; Goeman et al. 1990; Findlay et al. 1994). The fish community in Horseshoe Lake, Crow Wing County, was drastically influenced for more than 10 years by stocked pike (Anderson and Schupp 1986). Predation on 127- to 152-millimeter (5- to 6-inch) yellow perch nearly eliminated recruitment of perch to adult sizes and appeared to be the major factor causing a collapse of the perch population in Horseshoe Lake. Reduced growth rates and abundance of other fish (walleye and largemouth bass) were attributed to the reduction in perch. The void created by reduced perch numbers in Horseshoe Lake may have been filled by large numbers of small bluegill. Similarly, bluegill abundance increased and their growth rate declined following pike stocking in Murphy Flowage, Wisconsin (Snow 1974). In contrast, removing small (<600 millimeters) pike over 6 years, along with stocking some large pike and yellow perch in Hammal Lake, Aitkin County, was insufficient to restructure that fish community (Goeman and Spencer 1992).

Northern pike stocking in Grace Lake, Hubbard County, caused large declines in populations of yellow perch and white sucker, and the declines appeared to affect the walleye population (Wesloh and Olson 1962; Colby et al. 1987). A negative relationship between pike abundance and walleye abundance was also found in a large-scale statistical analysis of Minnesota lake surveys, although the negative effect was only clearly evident at very low pike abundances (Jacobson and Anderson 2007). Craig (1996) suggested that competition between pike and walleye is most likely to occur among young-of-the-year fish. In a 12-hectare (30-acre) German lake, stocking pikeperch *Sander lucioperca* on top of native pike and Eurasian perch *Perca fluviatilis* hardly affected the abundance, distribution, and consumption rates of pike, but the resulting higher levels of predation did cause changes in the way Eurasian perch used different habitats and in their abundance (Schulze et al. 2006). Following extirpation of pike and chemical treatments to reduce aquatic plants in Harriet Lake, an urban lake in Minneapolis, yellow perch and white sucker increased to high abundances, and bluegill abundance declined (Colby et al. 1987).

Several other studies provide intriguing glimpses into aquatic community dynamics after establishment of new northern pike populations. A large-scale reduction in yellow perch numbers, due to predation by pike stocked into Lake 221 of the Experimental Lakes Area in northwestern Ontario, caused changes that cascaded down to the microscopic zooplankton and phytoplankton communities in the lake (Findlay et al. 1994). After 7 years, a large proportion of the pike were removed from Lake 221, and yellow perch numbers rebounded to previous levels (Findlay et al. 2005). Invasion of pike into West Long Lake, Nebraska, led to reduced numbers but improved growth rates for bluegill, yellow perch, and largemouth bass within 4 years (Debates et al. 2003). Size structure of the bluegill and yellow perch populations in West Long Lake declined, apparently because pike predation had focused on the largest bluegill and perch. In contrast, size structure of the largemouth bass population increased owing to the additional predation pressure that pike put on smaller bass. Even the threat of pike predation can cause important behavioral changes in prey fish populations. Introduction of pike into a small bog pond in Wisall seasons. Furthermore, the roofs of pike mouths are lined with three wide strips of palatine teeth (Figure 1.38). The palatines slant inward toward the throat and make potent grips for holding both prey fish and human fingers. The bottom line is that at no time of year should you ever put your hand in a pike's mouth.

FIGURE 1.37. Canine teeth along the lower jaws of a northern pike.

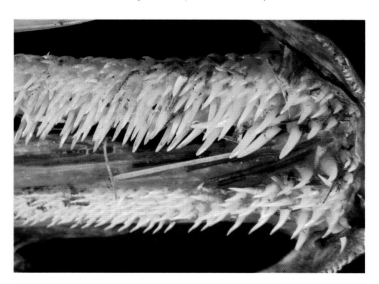

FIGURE 1.38. Palatine teeth of the upper mouth of a northern pike.

Chapter 2
Recreational and Commercial Fishing

Vulnerability of Northern Pike

Northern pike are valued principally as sport fish in Minnesota. One of the reasons for their popularity is that pike are very vulnerable to angling and are readily caught with spoons, spinners, lures, and bait (Figure 2.1). Pike fingerlings barely larger than 25 millimeters (1 inch) can be coaxed to nip at tinsel being pulled through the water. A study in Missouri (Weithman and Anderson 1978) stocked 20 small pike into man-made ponds during April. Every one of the pike was caught at least once between April and September, and one fish was caught seven times. In Shaw Lake, Michigan, a lake where the pike had never been exposed to fishing, two anglers reportedly caught 28% of the pike population in 11 hours of fishing (less than 2.5 angler-hours per acre) (Latta 1972). Three radio-tagged pike, whose locations were monitored in Potato Lake, northwestern Wisconsin, were readily caught by ice fishing (Margenau 1987). One of the fish was caught within 10 seconds after a hole was drilled in the ice and a jig (baited with a minnow) was dropped down toward the fish. Yet, pike have demonstrated they can learn to avoid being caught with artificial lures. Catchability to fishing with spinners declined to very low levels after about half of the population of pike stocked into a drainable pond in the Netherlands had been caught by anglers (Beukema 1970).

FIGURE 2.1. A commonly kept size of northern pike in the recreational fishery.

Fishing can have important effects on northern pike populations. In lakes where pike have good access to spawning and nursery habitat, overfishing generally leads to higher densities of small pike because longer-term exploitation can result in compensatory recruitment, where fishing actually stimulates the addition of new small pike to a population. During an 11-year period of intensive fishing in Heming Lake, Manitoba, the pike population continually increased with increasing fishing pressure while the average size of pike declined (Lawler 1965). In three Michigan lakes, Diana (1983b) found that the ages when pike matured corresponded with their mortality rates. Increasing levels of fishing intensity in the three lakes seemed to induce higher total mortality rates and earlier ages at first reproduction. Unexploited populations that are opened to fishing are quickly altered. The pike population of Brown's Lake, Iowa, was reduced by 85% in 4 weeks of intensive angling (Hill 1974). When Mid Lake, Wisconsin, was opened to fishing after being closed for more than 20 years, anglers harvested an estimated 46% of the naive pike population by the end of the first month, and much of the harvest occurred in the first 2 days (Goedde and Coble 1981). Experimental angling of an unexploited population in Savanne Lake, Ontario, removed 50% of the annual adult pike production even though angling effort was very low (1.2 hour per hectare [0.5 hour per acre]) (Mosindy et al. 1987).

Changes in age and size structure of fish populations are common symptoms of exploitation (Goedde and Coble 1981). Otto (1979) found that intensive fishing for northern pike in a small Swedish lake caused changes in size structure with an initial shift toward smaller fish. Historical records of gill netting in Lake Windermere, England, showed that large, old pike practically disappeared during the first two harvests on that lake, and younger age groups became relatively more important (Kipling and Frost 1970). Lawler (1965) found that the long-term intensive netting of pike in Heming Lake, Manitoba, resulted in a marked decrease in average size of the fish.

Creel Survey Statistics

Creel surveys are on-site studies collecting information about the numbers, sizes, and types of fish caught, as well as the fishing methods and number of hours of fishing effort expended in recreational fisheries. Creel clerks are assigned to work at the lake during specific time periods, and they monitor the fishing effort and interview fishers about their catches (Figure 2.2). Creel surveys give some indication of the success people can have angling for northern pike in lakes where the fish have been continuously exposed to fishing pressure. Creel surveys from seven north-central Minnesota lakes (during 1988–91) found a typical angling catch rate of one pike for every 5 hours of fishing (0.2 fish per hour) during the summer, and that anglers released 75% of the fish they caught (Pierce et al. 1995). These catch rates applied to everyone who was fishing, although anglers who were specifically fishing for pike had higher catch rates, which averaged about 1 fish per 2 hours (0.5 fish per hour). In winter, both catch rates and release rates for anglers were nearly 30% lower than in summer. Similarly, creel surveys from 55 lakes in northern Wisconsin (during 1990–99) found average catch rates of 0.3 fish per hour for open-water anglers, and 0.2 fish per hour for ice anglers specifically fishing for pike (Margenau et al. 2003). The rate at which people released pike they had caught exceeded 80% during the open-water fishery in northern Wisconsin but was less than 50% for the ice fishery. Resulting harvest rates were 0.06 fish per hour during open water, and 0.11 fish per hour for ice fishing, implying that there is much more interest in keeping pike for food during the ice fishery than during open-water fishing. Most pike harvested from the Wisconsin lakes were 457 to 607 millimeters (18.0 to 23.9 inches) total length.

A large-scale historical analysis of Minnesota creel surveys (Cook and Younk 1998) indicated that anglers released 62% of the northern pike they caught during the summer, compared with anglers in winter, who released 26% of their pike catch. Average

FIGURE 2.2. A creel clerk interviews anglers as they return from a fishing trip.

lengths of harvested pike in Minnesota were 21.8 inches (554 millimeters) during open water and 22.6 inches (574 millimeters) in winter, in contrast to average lengths of released pike that were 17.6 inches (447 millimeters) during open water and 18.6 inches (472 millimeters) in winter. Over 90% of fish that were kept by anglers were between 16 to 29 inches (406 and 737 millimeters) long. The Minnesota analysis concluded that fish over 24 inches (610 millimeters) have seldom been released, whereas over 85% of the fish that are released are less than 20 inches (508 millimeters). A graph of the cumulative harvest in relation to pike lengths (Figure 2.3) further illustrates that the bulk of the harvest is over 20 inches (508 millimeters) in Minnesota waters.

Although people have been unlikely to release large northern pike in the recent past, the growing popularity of catch-and-release fishing, along with an increasing awareness of the value of large pike, may lead to higher release rates for large pike in the future.

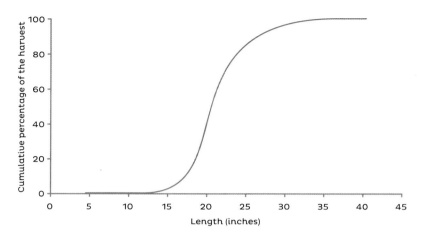

FIGURE 2.3. Cumulative harvests in relation to the length of northern pike.

Hooking Mortality

Mortality rates of northern pike that are caught and released can be kept relatively low, especially if the fish are not deeply hooked and are handled carefully. A notable exception is when fish are caught with pike hooks (also known as Swedish hooks) during ice

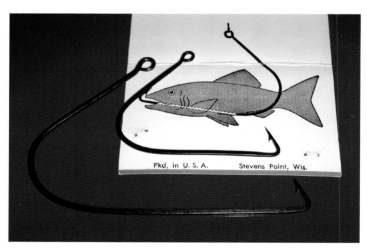

FIGURE 2.4. Pike hooks cause high hooking mortality rates.

fishing (Figure 2.4). Pike hooks baited with dead rainbow smelt *Osmerus mordax* and fished from tip-ups caused substantial mortality (33%) due to deeper hooking and more extensive bleeding (Dubois et al. 1994). In contrast, the same study found that hooking mortality was negligible (less than 1%) when pike were caught on treble hooks baited with live fish.

A review of the scientific literature on hooking mortality for northern pike (Tomcko 1997) found that hooking mortality averaged 4.5% among six studies (mortality from pike hooks was excluded). The individual studies covered various fishing methods and reported mortality rates of 0% to 14%. Pike had greater physical damage and mortality rates if baits were swallowed and the fish deeply hooked, a result commonly documented in hooking mortality studies for other fish species. A more recent study of hooking mortality in Canadian and German lakes (Arlinghaus et al. 2008) confirmed those results for pike and demonstrated that hooking in critical locations, such as the gills and gullet, was more likely to occur with natural bait, soft plastic lures, and jigs. Radio-tracking studies with pike in a German lake showed that sublethal catch-and-release impacts on behaviors, such as movement and habitat choice, were short-term and reversible (Klefoth et al. 2008; Arlinghaus et al. 2009).

Darkhouse Spearing

Darkhouse spearing through the ice is a traditional form of northern pike harvest during Minnesota winters. The origins of pike spearing can undoubtedly be traced back to Native American winter fishing techniques and the tendency of early lawmakers and fishers to classify pike as rough fish. The technique involves cutting a large hole in the ice and suspending live bait or artificial decoys in the water to attract pike within spearing distance. Low light levels in a darkhouse allow the spearer to see down into the water column (Figure 2.5). An excellent treatise on the methods used for spearing is the book *Darkhouse Spearfishing across North America* written by Jay Leitch (1992). Minnesota is not unique in allowing winter spear fishing; five other states allow at least limited spear seasons, although no other state has experienced the amount of effort expended by spearers in Minnesota (Leitch 1992) (Figure 2.6).

The legacy of spearing in Minnesota is long and laced with controversy. Various groups have charged that spearing is not compatible with other uses of the fisheries. Meanwhile, darkhouse spearers have fought efforts to curtail their sport with organized political efforts. As early as 1929, some lakes within Minnesota were closed to spearing, and legislation in 1941 permitted closure of up to 50% of the state's waters to spearing. In 1947, when the commissioner for the Conservation Department (now the Department of Natural Resources) cut the spearing season from 77 to 30

FIGURE 2.5. A small northern pike approaches the white sucker tethered as a decoy beneath a darkhouse, as viewed through the spearing hole in the floor of the darkhouse.
PHOTOGRAPH COURTESY OF SAM JOHNSON.

FIGURE 2.6. Itasca County spearers.
PHOTOGRAPH COURTESY OF SAM JOHNSON.

FIGURE 2.7. Newspaper clippings from the 1950s about spearing issues.

days, darkhouse groups organized large-scale protests (Figure 2.7). The ire of spearers peaked during the years from 1947 to 1953 over the length of the winter spearing season and over controversial closings of lakes to spearing (Pierce 1998). Meanwhile, spearing reached the height of its popularity in the 1950s, and the first spearing licenses were issued in 1955.

Conflicts between spearers and anglers, particularly about harvests of large northern pike, have led to questions about relative harvests by each group and their effects on pike populations (Latta 1972). Minnesota's statewide creel survey database has shown that spearers harvest pike at a rate similar to summer and winter anglers who are specifically fishing for pike. Long-term average harvest rates have been 0.175 fish per hour for spearing, 0.185 fish per hour for summer angling, and 0.193 fish per hour for winter angling (Pierce and Cook 2000). Because there are fewer spearers than anglers, spearing has clearly accounted for fewer fish harvested than angling. Spearing has averaged 15% of the total pike yield by number, and 22% of the total yield by weight (Pierce and Cook 2000).

Several investigations have found that spearers take larger fish than anglers (Johnson et al. 1957; Schupp 1981; Diedrich 1992). In an individual lake, this was illustrated during 3 years in which both summer and winter creel surveys were obtained. The average weight of pike speared in Ball Club, a 1,593-hectare (3,936-acre) lake in Itasca County, was consistently greater than the average weight caught by anglers during the 3-year study (Johnson and Peterson 1955). Pike length measurements and ages of fish recorded in the statewide creel survey database also illustrate that the spearing harvest contains a greater proportion of larger-size, older fish than the angling harvests (Figure 2.8). Ages from harvested pike (Figure 2.8) show that spearing harvests peak for age-3 pike. Fish of ages 2 to 5 comprised more than 84% of the harvest in nine surveys for which ages of the speared pike were determined (Pierce and Cook 2000).

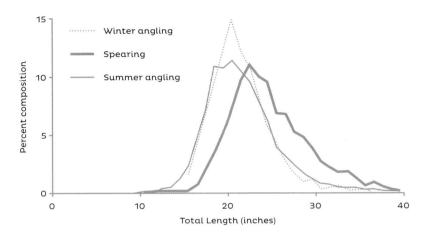

FIGURE 2.8. Percentages by length and age of northern pike harvested by spearing and by summer and winter angling in Minnesota lakes as determined from creel surveys. Age data were not available for winter angling (Pierce and Cook 2000).

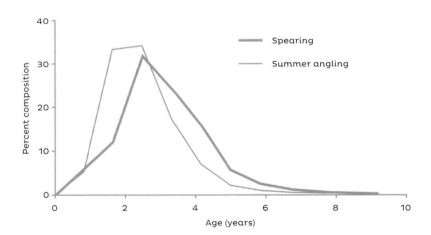

Northern Pike as Food for People

Special filleting techniques are required when cleaning northern pike to remove the obnoxious floating "y-bones," which detract from edibility. One popular approach is to cut 3-inch steaks, then fillet out the ribs and y-bones. In his book *Fixin' Fish*, Gunderson

(1984) provides pictures illustrating an alternate set of special cuts for filleting pike. The cuts include a fillet along the back, two simple fillets near the tail, and two more intricate cuts along the y-bones in the midsection of the fish. Gunderson's book also contains a recipe for pickling, which is a delicious and commonly used method for preserving pike meat in a vinegar preparation. Properly simmering or freezing the flesh is important, however, because the broad fish tapeworm *Diphyllobothrium latum* is a parasite acquired by eating raw or improperly cooked northern pike. The adult tapeworm develops in the intestines of humans and other mammals, growing up to 9 meters (30 feet) long and persisting for as long as 25 years when untreated (Pullen et al. 1981).

A commonly observed parasite is "black spot," which refers to a pepper-like sprinkling of black spots on the skin and fins of pike and other fish (Figure 2.9). Black spot is an encysted stage of parasitic flatworms *Neascus* spp. These parasites have a complex life cycle using a series of host organisms that include snails and fish-eating birds. Black spot is not known to infect humans but has potential effects on growth rates and mortality of pike. In the Niagara River, New York, infected pike were smaller than uninfected pike at ages 1 to 4, and the size difference increased with age (Harrison and Hadley 1982). Moreover, incidence of infection decreased with fish size in the Niagara River, and no pike over 600 millimeters had the distinctive black spots. All parasites that humans can get from pike are killed by thoroughly cooking the meat.

As a top-level predator, northern pike tend to bioaccumulate pollutants such as mercury and polychlorinated biphenyls (PCBs) to a higher concentration than fish lower on the food chain, and the largest fish accumulate the highest concentration of contaminants. Mercury, in particular, is widespread in Minnesota lakes. High levels of mercury contamination in pike are not natural and are the end result of mercury pollution of the atmosphere. Mercury in the air can be transported long distances before it

FIGURE 2.9. Characteristic black spots from the encysted stage of *Neascus*.
FILE PHOTOGRAPH FROM MINNESOTA DEPARTMENT OF NATURAL RESOURCES.

is washed out by rain and falls to the landscape. The Minnesota Department of Health provides guidelines about health risks of eating contaminated fish (www.health.state.mn.us/divs/eh/fish/). The department advises limiting meals of pike for adults when mercury concentrations exceed 0.16 parts per million. The threshold is even lower (0.05 parts per million) for young children and women who are breastfeeding or who can become pregnant.

Mercury concentrations in pike in Minnesota tend to be higher in the northeastern part of the state, where mercury concentrations in a few lakes exceed 0.6 parts per million in a standardized size of fish (550 millimeters [21.7 inches] total length). While atmospheric deposition of mercury is relatively uniform across northeastern Minnesota, contamination in the fish varies widely among lakes because of differing limnological and watershed characteristics that influence mercury uptake (Sorensen et al. 1990; Wiener et al. 2006). For example, among lakes located on the Kabetogama Peninsula (an area within Voyageurs National Park that is largely unaffected by human development), mercury concentrations in northern pike vary by an order of magnitude, from less than 0.2 parts per million to 2.3 parts per million for a standard-size pike (data from G. E. Glass and the Minnesota Pollution Control Agency; also see Sorensen et al. 1990 for similar information). An example demonstrating that the largest fish accumulate the greatest amount of contaminants is Sand Point Lake, St. Louis County, where mercury concentrations were less than 0.3 parts per million for pike smaller than 450 millimeters (17.7 inches) but increased progressively to concentrations that sometimes exceeded 1.5 parts per million for pike longer than 700 millimeters (27.6 inches) (data from G. E. Glass and the Minnesota Pollution Control Agency; also see Sorensen et al. 1990).

A deliberate reduction of mercury concentrations was attempted by overfishing pike in a small Finnish lake (Verta 1990). Intensive overfishing reduced the mercury levels in large pike, but contamination apparently increased in young pike, a result that might be explained by an earlier switch from an invertebrate to a fish diet. While intensive fishing was estimated to have removed several years' accumulation of methylmercury in the lake's fish, fishing obviously did not address the root problem of continuing mercury pollution.

An unsettling trend has been increasing levels of mercury in pike in Minnesota since the mid-1990s. Evaluation of fish mercury concentrations over a 25-year period (1982-2006) indicated a downward trend until the early to mid-1990s, and an upward trend thereafter (Monson 2009). The upward trend in pike mercury concentrations is difficult to attribute to any one factor, but may be a response to earlier trends in global emissions of mercury combined with climate change (Monson 2009).

Time Trends in Minnesota's Northern Pike Fisheries

Trends in the Recreational Fishery

In Minnesota, separate licenses are required for spearing, angling, and placing portable or permanent houses (shanties) on the ice. The numbers of angling and icehouse licenses sold have trended upward in time as human populations and interest in fishing have expanded (Olson and Cunningham 1989; Cook et al. 1997). In the early 1930s, fewer than 500,000 angling licenses were sold. With the exception of a period during World War II, license sales increased rapidly from the 1930s through the 1960s, and by the year 2000, the number of licensed anglers (both resident and nonresident) exceeded 1,580,000. Historical icehouse license sales increased from an average of 50,526 licenses per year during 1949-59 to an average of 142,509 per year during 1999-2004. In contrast to the increases in angler and icehouse licenses, darkhouse spearing license sales documented the decline of a unique sport fishery for northern pike. As evidenced from license sales, the heyday of spearing in Minnesota was in the 1950s. Since then, there has been

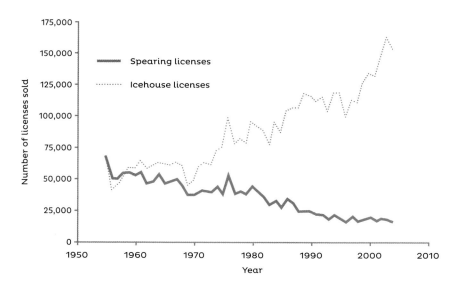

FIGURE 2.10. Historical numbers of spearing and icehouse licenses sold in Minnesota.

a long and relatively consistent decline in darkhouse spearing for pike. License sales progressively decreased from 64,833 licenses sold during 1955 (when spearing licenses were first sold) to an average of 17,540 licenses per year during 1999–2004 (Figure 2.10).

Reductions in effort and harvests by darkhouse spearers have mirrored the decline in license sales (Pierce and Cook 2000). The decline in spear-fishing activity cannot be blamed on catch rates, however, because spearing catch rates have not changed over time (Pierce and Cook 2000). Rather, the tradition of spearing has not been as readily adopted by current generations of sport fishers. Most spearers indicate that spearing is a tradition learned as a youngster from their fathers or grandfathers (Borge and Leitch 1988). Although the number of spearers has been greatly reduced, the remaining core group of spearers is active, and some members belong to the Minnesota Darkhouse and Angling Association, an organization created to preserve their sport. As a result of the comparatively low number of spearers and their devotion to their sport, the decline in spearing license sales has leveled off since

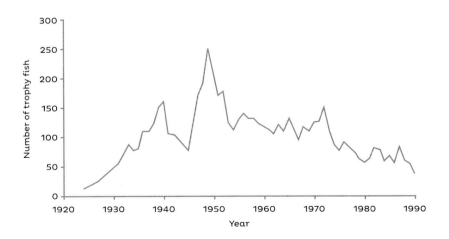

FIGURE 2.11. Numbers of trophy northern pike entered each year from 1924 to 1990 in Fuller's Tackle Shop fishing contest, Park Rapids area of northwest Minnesota.
DATA WERE PROVIDED BY PETER JACOBSON (MINNESOTA DEPARTMENT OF NATURAL RESOURCES, DETROIT LAKES).

about 1995 (Figure 2.10). Meanwhile, increases in angling pressure have far surpassed the reductions in spearing effort (Pierce and Cook 2000).

Sizes of fish have suffered from the historical increases in fishing effort in Minnesota. A unique analysis of long-term records from an annual fishing contest in the Park Rapids region of northwestern Minnesota offered insights into historical changes in population size structure of northern pike in response to increasing levels of exploitation by recreational fishing (Olson and Cunningham 1989). Records published annually from 1924 to 1990 were available for the fishing contest of Fuller's Tackle Shop in Park Rapids, Hubbard County. The average weight of individual pike entered in the contest declined from 4.6 kilograms (10.1 pounds) in the 1930s to 3.1 kilograms (6.8 pounds) in the 1980s. Figure 2.11 illustrates annual changes in numbers of trophy-size fish that were entered in the Fuller's Tackle Shop's fishing contest, where trophy pike were defined as fish weighing at least 10 pounds (4.54 kilograms). Under increasing exploitation, numbers of trophy-size fish increased through the 1920s and 1930s. Numbers declined with reduced fishing effort during World War II, then spiked rapidly to a peak in 1948 as interest in recreational fishing expanded after the war. Since then, numbers of trophy-size fish entered in the contest have steadily declined, seeming to confirm that fishing nowadays is not the same as the "good ole days."

Commercial Fishing Records

Northern pike were harvested commercially from the Mississippi River and Minnesota's border waters with Canada, with the bulk of the harvests occurring in the early 1900s. The tremendous expansion of sport fishing and its economic value led to the demise of commercial fishing. Commercial catches of pike estimated for all of the states along the Upper Mississippi River decreased from 50,800 kilograms (112,000 pounds) in 1894 to about 2,300 kilograms (5,000 pounds) by 1931 (Carlander 1954). Records of commercial harvests of pike in Canadian border waters date back to 1888 for Lake of the Woods, Lake of the Woods and Roseau Counties; 1908 for Rainy Lake, St. Louis and Koochiching Counties; and 1919 for Namakan Lake, St. Louis County. By 1920, 127 licenses were issued to commercial fishers, who took pike from those three waters along with Lake Kabetogama, St. Louis and Koochiching Counties. Commercial fishing for game fish was phased out in 1926 in Lake Kabetogama because of the growth of the tourism industry and the influence of resorts that based some of their reputations on walleye fishing. In Namakan Lake, commercial harvest of pike was allowed until 1946. The highest annual commercial harvest of pike in Namakan Lake was 47,068 kilograms (103,765 pounds) in 1929. Annual pike harvests from 1919 to 1935 in Namakan Lake averaged 13,510 kilograms per year (29,783 pounds per year) but declined to 1,236 kilograms per year (2,724 pounds per year) from 1936 to 1946. The cumulative recorded pike harvest from Namakan Lake from 1919 through 1946 was 243,251 kilograms (536,268 pounds).

By comparison, the cumulated total recorded northern pike harvest for Minnesota waters of Rainy Lake was nearly 0.8 million kilograms (1.8 million pounds; records from 1908-84). The highest annual commercial harvest was 39,836 kilograms (87,819 pounds) in 1920, but commercial harvests of pike also declined over time in Rainy Lake. Annual harvests averaged 28,970 kilograms per year (63,867 pounds per year) during 1908-35, and 9,510 kilograms per year (20,966 pounds per year) during 1936-55, and dropped to 3,555 kilograms per year (7,837 pounds per year) during 1956-84. The 1984 catch of 1,912 kilograms (4,216 pounds) was valued at $1,360 ($0.71 per kilogram [$0.32 per pound]). Commercial game fish licenses and quotas were purchased in a "buy-out" by the state in 1985.

The largest commercial harvests of northern pike in Minnesota were from Lake of the Woods. Records of commercial fishing in Lake of the Woods from 1888 through 1985 provide a minimum estimate (records are missing for 9 years) of over 5 million kilograms (11 million pounds) of pike removed from Minnesota waters of the lake during the 98-year period of commercial fishing. The highest recorded catch was 238,140 kilograms (525,000 pounds) during the 1915 commercial fishing season. Catches during 1915-35 averaged 116,211 kilograms per year (256,196 pounds per year). During the last 35 years of commercial fishing (1950-84), catches averaged 30,340 kilograms per year (66,888 pounds per year). The Lake of the Woods commercial fishery was also "bought out" by the state during 1985-86.

Chapter 3

Conservation and Management of Northern Pike

Conservation and Management Strategies in Minnesota

Current Conservation Issues

Loss of critical habitat has been an important issue for maintaining northern pike populations (Casselman and Lewis 1996; Margenau et al. 2008). Draining and filling of wetlands and so-called "improvement" of shorelines for lake homes have been increasingly responsible for lost habitat in urban, agricultural, and other highly developed areas of Minnesota. Shoreline and related land development removes vegetation, reduces water quality, and reduces dissolved oxygen levels in the sediments (Burns 1991; Cross and McInerny 1995; Radomski and Goeman 2001). A study of habitat used by young-of-the-year pike in Spirit Lake, Iowa (Bryan and Scarnecchia 1992), found pike only along natural shorelines; the fish avoided developed lakeshore. An essay by Clifford Brynildson titled "What's Happening to Northern Pike Spawning Grounds?" published in the *Wisconsin Conservation Bulletin* in 1958, documents that we have been aware of the problem of habitat loss for at least half a century. Yet, shoreland zoning regulations, which were first adopted in 1970, have failed to stem the loss of habitat. Where habitat has been altered in southern Minnesota, stocking is a last resort for maintaining pike populations. Fry are stocked directly from hatcheries, or fry or gravid (egg-bearing) adults are stocked into managed wetlands, where the young pike are allowed to grow for a few weeks to small fingerling size before being released into adjoining lakes. Future habitat alteration may also include changes in thermal habitat (and even prey species associations) related to global climate change. Such responses have already been detected in Lake Windermere, England, since the 1990s (Winfield et al. 2008).

Where good natural habitat for northern pike exists, natural reproduction is usually not a limiting factor. In fact, a common phenomenon in many small central and northern Minnesota lakes is large numbers of small, stunted (slow-growing) pike. From a fisheries management viewpoint, these populations are difficult to manage because they arise from some combination of overharvest of large fish, a lack of appropriate-sized prey fish, and habitat characteristics that fail to promote good growth (Pierce et al. 1995; Paukert et al. 2001). Maintaining an appropriate balance of large pike, in the face of heavy fishing pressure on large fish, may be a key problem for managing pike populations. Stocking has been an unsatisfactory management tool in such lakes and has only worsened problems with prey fish communities and poor growth rates of pike.

The Evolution of Management Strategies in Minnesota

The historical progression of management techniques for northern pike in Minnesota is an interesting story of changing perspectives in conservation. Early in the history of the state, when human population densities were low, pike were abundant enough to be considered rough fish along with suckers and bullheads. Minnesota's Fish Commission, in its first annual report in 1874, regarded the pike as a "calamity" of nature, and the commissioners were "fully convinced that every pickerel of the state simply occupies the room of a better fish" (Hoffbeck 2001). As human densities began to increase, pike became more valued, and it was recognized that they might not fare well under unfettered exploitation. The earliest management, then, consisted of regulations that restricted techniques for catching pike, closed down fishing during spring spawning periods, and began to limit people from harvesting overly large numbers of fish in a single day.

Meanwhile, climatic conditions were changing in Minnesota. Amounts of rainfall that the state received declined during the early 1900s and were very low during the drought years of the 1930s. Similarly, lake levels and runoff declined from 1920 to 1940. We cannot document what happened to pike populations, but it is likely that low water levels led to relatively low pike population numbers. During this period, attempts were made to take pike eggs, incubate them in hatcheries, and stock the fry directly into lakes or into holding ponds, where they were raised to fingerling sizes. Early propagation efforts were not very successful, because the eggs were difficult to handle and because yields from the ponds were low as a result of the cannibalistic nature of pike. The average annual yield from rearing ponds in the 1940s was 12 pounds of fingerlings per acre (13.4 kilograms per hectare) (Moyle 1949).

A sharp rise in recreational fishing activity occurred after World War II and stimulated more active management techniques for northern pike. Water levels also influenced the new developments in pike management over the two decades following World War II. A veteran state fisheries biologist, Dennis Schupp, suggested that late and dry winters followed by low lake levels in spring during the late 1950s reduced natural recruitment. This observation prompted more intensive pike management, and imaginative and productive methods for raising and stocking pike evolved quickly. One of the earliest of those methods was "winter rescue." Young pike and adults that had moved into lakes prone to winterkill were "rescued" from the winterkill lakes by unique trapping techniques in late winter. Water-pumping stations generated aerated water that attracted pike into artificial channels, where they were trapped. Fish caught during winter rescue operations

FIGURE 3.1. A northern pike of 22.5 pounds speared from Big Splithand Lake, Itasca County, January 1949.
PHOTOGRAPH COURTESY OF THE ITASCA COUNTY HISTORICAL SOCIETY PHOTO COLLECTION.

were stocked in large numbers throughout Minnesota. The numbers of pike that were rescued increased through the 1970s and then declined after water levels rose and the technique began to fall out of favor with fisheries managers.

Another new technique in the fish manager's arsenal was actively managed spawning areas that included both natural sloughs and constructed ponds. Natural movement of spawning northern pike into the sloughs or active stocking of mature pike into the ponds allowed spawning to take place. Control structures maintained high water levels after spawning until the developing fingerlings could be released. Managed spawning areas were viewed as a technique to counter increasingly intensive lakeshore development and draining of wetland habitat.

Northern pike abundance in Minnesota increased rapidly in the early 1960s and remained at levels substantially higher than those observed from 1948 through 1960 (Schupp 1981). It is plausible that higher spring water levels coupled with intensified pike management (winter rescue and managed spawning areas) from the preceding 20 years both contributed to the increase. After the 1970s, gradually reduced stocking from winter rescue and reduced production from managed spawning areas reflected a lessening need for managers to build low populations compared with earlier years when water levels were less favorable for natural reproduction (Schupp 1981). Recognition was also growing about potential fish community problems resulting from stocking a top-level predator like the pike (Maloney and Schupp 1977; Anderson and Schupp 1986).

With higher water and greater levels of natural reproduction, active management of northern pike was largely dropped in Minnesota, and we entered a period during the late 1980s and 1990s when pike were mostly left on their own to reproduce and prosper. However, in the 1980s, a vocal set of anglers became increasingly concerned about the small sizes of pike in many lakes where natural reproduction was more than adequate, and wanted to know whether fish sizes could be improved in these "hammer-handle" lakes. The MNDNR responded by implementing experimental regulations in 20 lakes around the state. Experimental regulations liberalizing bag limits to six small pike were initiated during the middle to late 1980s. Expanded bag limits were an attempt to reduce numbers of pike by allowing people to keep more small pike from high-density, slow-growing populations. Unfortunately, creel surveys showed that anglers simply did not use the expanded bag limits even after they claimed they would. Less than 1% of angling parties took advantage of the expanded limits (Pierce and Tomcko 1997), and further evaluations concluded that anglers would be hard-pressed to remove enough small pike to effect meaningful changes in the fish populations (Goeman et al. 1993). Since the expanded bag limits were largely ineffective for promoting harvest of small pike, they were dropped.

Yet concerns about small sizes of pike continued, and length limits were considered as an alternative approach. Experimental catch-and-release and minimum, maximum, and slot length limits were initiated in 25 lakes from 1989 through 1998. Many of these experimental regulations were in place for 9 to 15 years. Evaluations of the regulations are discussed in the section "Evaluation of Special Regulations" later in this chapter.

During the 1990s, interest continued to build among the fishing public for more extensive use of size limits in managing northern pike. Ever-increasing numbers of people had a sense that there had been long-term declines in fish sizes, that there were too few opportunities to catch big pike, and that length limits were a potential tool for dealing with the problem. In 1999, the MNDNR began a review of fishing regulations for all fish species, a review that resulted in a proposal for a statewide slot length limit protecting pike between 24 and 40 inches long. During the ensuing public meetings, however, it became apparent that a large majority of the fishing public was more supportive of managing pike on a lake-by-lake basis rather than with a statewide limit, so the idea of a statewide length limit was dropped. By 2002, the list of lakes with special regulations for pike had grown to 32, and public pressure continued to mount for improving pike sizes in more lakes. The MNDNR once again responded by initiating special regulations at 76 additional lakes beginning in 2003. Regulations were picked from a simplified "toolbox" of three regulations: a 40-inch minimum length limit designed to promote trophy fisheries, a slot length limit protecting 24- to 36-inch fish intended to improve sizes of fish in lakes with good natural reproduction, and a 30-inch minimum length limit for lakes with poorer natural reproduction. Formal evaluations of the toolbox regulations will not begin until 2013.

In retrospect, the history of northern pike management in Minnesota can be viewed as a progression from early regulations followed successively by hatchery production, winter rescue and managed spawning areas, "hands-off" management, and then special regulations emphasizing individual lake management. This evolution of management strategies was influenced primarily by increases in fishing pressure, historical changes in natural water levels, shoreland development that altered natural habitat, and shifts in how pike were valued as a recreational and ecological resource. Historical management techniques that have been used for pike are described in more detail in the following sections.

Statewide Fishing Regulations
Early Attempts at Regulation

Early lawmakers classified northern pike as rough fish along with suckers and bullheads so that harvests were unregulated prior to 1864. In 1864, an "Act for the Preservation of Elk, Deer, Birds, and Fish" restricted legal methods for taking fish to shooting with a gun, hook-and-line, and spearing. Legislation in 1891 authorized hiring of a superintendent of fisheries and construction of fish hatcheries within Minnesota. During these earliest years of regulation, the debate was about appropriate methods and times of year for taking fish and produced a confused flux of regulations. The 1891 act prevented wanton waste of fish, closed the fishing season during March and April, and limited harvest methods for recreational fishing to angling with hook-and-line or spearing. An amendment in 1893 declared that fish were the sole property of the state and authorized a closed season from December 1 to April 15, which effectively eliminated ice fishing. The closed winter season was apparently unpopular, because the closed season reverted back to March through April in 1895. Beginning in 1897, spearing was allowed year-round, and a 6-inch minimum size limit was established for all fish, but ice-fishing houses were banned. Use of artificial lights while fishing was legalized in 1905 but subsequently restricted during May and June in 1907. All of these early regulations in Minnesota referred to northern pike as "pickerel" or "great northern pike," whereas the term "pike" referred to walleye.

The early Board of Game and Fish Commissioners was abolished in 1915 and replaced with a single commissioner appointed by the governor. This important restructuring led to new regulations in 1919 that were more specific for northern pike, setting a 14-inch minimum size limit, a 25-fish daily bag limit, and a closed season during early spring (March–April) for both angling and spearing. Subsequent regulations have served mainly to modify these limits.

Statewide Length and Bag Limits

The minimum length limit of 14 inches imposed in 1919 was in effect through 1930. A 16-inch minimum length limit replaced it in 1931 but was quickly removed in 1932, and there have been no statewide size limits for northern pike since 1931. From 1930 through 1938, possession limits were 20 pike, and daily bag limits were 10 fish. The daily limit was lowered to 8 pike beginning in 1939, and the possession limit lowered to 12 fish in 1941. Bag limits quickly became more conservative, dropping to 6 fish (possession and daily limit) in 1947, then to 3 fish in 1948. The changes reflected

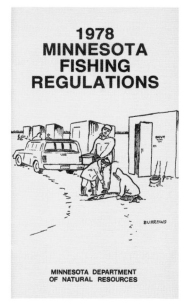

FIGURE 3.2. Covers from Minnesota's fishing regulations booklets.

the impacts that fishing was having on pike populations, because the rather quick reduction in bag limits coincided with a period when Olson and Cunningham (1989) noted that average weights and numbers of trophy-sized pike began to decline. Possession and daily bag limits remained unchanged from 1948 until 1994 when statewide bag limits were further restricted by the legislature to allow only 1 fish over 30 inches as part of the 3-fish limit. The 1-over-30 regulation was not proposed by the MNDNR but, rather, was a popular initiative from the legislature. In retrospect, it is evident that historical reductions in Minnesota's bag limits have not curtailed the decline in fish sizes over the past 60 years.

Open Seasons

Changes in the fishing season have consisted of tinkering with the dates to close fishing in late winter under the notion that northern pike should be protected while they are staging for spawning, a time when they might be especially vulnerable. From 1922 through 1933, open seasons were May 15 through February 1. The winter season was extended to March 1 from 1934 through 1939, and then reverted to mid-February in 1940. Open angling seasons were consistently mid-May through mid-February until 2006. After 1930, spearing was relegated to the winter only, and most seasons began on December 1. With a few controversial exceptions, listed below, the statewide closing date for spearing was mid-February from 1945 through 2006. In 2006, the open season for angling was once again extended to the end of February, and the following year the open spearing season was also extended to the end of February.

The length of the winter spearing season has generated many disputes, and much of the controversy was centered around Commissioner Chester Wilson (Itasca County Darkhouse Fishing Association et al. 1954). Darkhouse spearing groups organized protests when the 1948 season was cut from 77 to 31 days (December 1-31). Additional shortened seasons from 1950 through 1952 prompted a roundtable discussion between the commissioner and darkhouse groups in Grand Rapids in March 1952, moderated by Duluth judge

ated strategies for enforcing pike length limits (Walker et al. 2007). Anglers in the Alberta study perceived that their chances of getting caught violating length limits increased as law enforcement effort increased up to the point where game wardens had contacted at least 3% of the anglers.

Length regulations for northern pike are intended to preserve or enhance the sizes of fish in a population by protecting moderate- to large-sized fish, which seem to be very susceptible to overharvest. The evaluations described above have shown the range and magnitude of responses we can reasonably expect from length limits, including the possibility that there will be no change in fish sizes. The results have shown that length regulations are among the most promising tools for managing pike populations, but the regulations result in an important and controversial trade-off. The trade-off is that length regulations reduce the total amounts of fish harvested to gain larger-sized fish for the recreational fisheries. The overall message from long-term evaluations of length regulations in Minnesota is that there is a substantial value to conserving large pike when our goal is to improve the sizes of pike (Pierce 2010b).

Liberalized Bag Limits

Liberalized bag limits for northern pike were implemented experimentally at 20 lakes in north-central Minnesota between 1986 and 1991. The liberalized bag limits were intended to improve the sizes of pike in "hammer-handle" populations by promoting harvest of small pike in excess of the general statewide limit of three fish. The liberalized bag limits were six fish, but to emphasize harvest of small fish, only three fish were allowed over 22 or 24 inches long in most of the lakes. Available data from 10 lakes (Gull, Sallie, Melissa, Detroit, Dead, Star, Birch, Pleasant, George, and Nord located in Aitkin, Becker, Beltrami, Cass, Hubbard, and Otter Tail Counties) showed that liberalized bag limits were generally not successful in improving pike sizes. Improved pike sizes were observed only in

Sallie, Melissa, and Star Lakes and may have resulted from natural variation in recruitment of small pike into the populations rather than the bag limits (MNDNR file data).

An analysis of intensive trap netting to remove small northern pike from Hammal Lake, Aitkin County (Goeman et al. 1993) showed that intensive trapping did not improve the sizes of pike, and creel surveys projected that anglers harvest substantially fewer pike than can be removed by intensive trapping. Goeman et al. (1993) concluded that even if anglers could be induced to remove more small pike, harvest levels would still not be sufficient to increase pike growth rates and to increase the proportion of large fish in such populations. An important conclusion reached after reviewing all of Minnesota's trials with experimental regulations is that protection of large pike with length limits has a stronger conservation value than liberalized bag limits. Liberalized bag limits convey a mixed message about managing large predatory fish to improve fishing quality. In many lakes, we have probably fished our way into poor size structure by overharvesting large pike, and we do not seem to be able to fish our way back out of that predicament by harvesting more small pike.

Hatchery Production

Hatchery production of northern pike in Minnesota is very limited compared with that of other states and compared with production of other species such as walleye. Modern pike culture at the Waterville hatchery began in 1991, and from 1991 through 2006, egg take at the Waterville hatchery averaged only 2.5 million eggs each year with an average annual production of 1.1 million fry for stocking. Cost for the Waterville operation during that time averaged $5.86 per 1,000 fry produced. The cost estimate was comprehensive: it included labor for collecting brood fish and incubating the eggs, electricity and water-pumping costs, vehicle mileage, and some maintenance and other miscellaneous costs.

Figures 3.8 through 3.10 illustrate current methods used for fertilizing and hatching northern pike eggs (developed by Bruce Pittman) at the Waterville hatchery. These methods replaced earlier dry fertilization methods that relied on semen extender. First, about 3 to 6 cubic centimeters of milt are aspirated from several males into a glass vial. A 1.3% baking soda solution (49 grams sodium bicarbonate dissolved in 3.8 liters of water) is poured into a ceramic pan until the solution is approximately 25 to 50 millimeters deep, and eggs from one or more females are stripped into the pan. Milt is added to the eggs, and they are gently swirled and stirred for 1 to 2 minutes. The very fragile fertilized eggs are gently rinsed into fine-mesh holding trays in hatchery raceways, where the eggs are allowed to harden overnight. Hardened eggs are poured into "Pennsylvania" hatching jars at a rate of 1 to 2 liters of eggs per jar, with a liter equivalent to 51,000 to 70,000 eggs depending on the size of the eggs. Water flow through the jars is

FIGURE 3.8. Aspirating milt from a male northern pike.

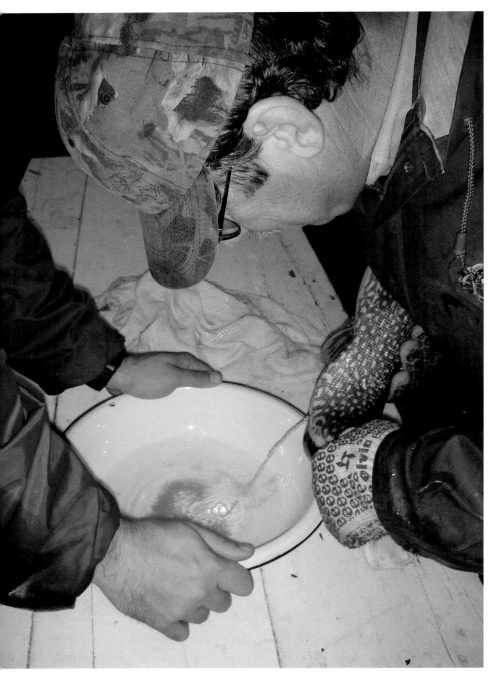

FIGURE 3.9. Stripping eggs into a ceramic pan, then gently stirring the eggs, milt, and baking soda solution with a feather.

FIGURE 3.10. Rinsing fertilized eggs into a holding tray, adding hardened eggs to a Pennsylvania jar, and jars of eggs on the battery.

very low for the first 5 days to prevent the eggs from tumbling or rolling until the embryos are well developed. Approximately 180 temperature units are needed to hatch the eggs, and another 150 units are necessary for the fry to reach an active swimming and feeding stage. Temperature units are calculated as the number of degrees Fahrenheit (°F) the water temperature is above 32°F, multiplied by number of days. Thus, if water temperature averaged 50°F for one day, that would be equivalent to 18 temperature units. Newly hatched fry are retained in the jars for about 4 to 5 days (by clamping screens down on top of the jars) until their yolk sacs are mostly expended and they become active. After the fry are actively swimming and feeding, they are stocked into lakes or rearing areas. Depending on water temperatures, the whole process from egg stripping until the fry are stocked lasts 14 to 20 days.

Some records document northern pike fry stocking prior to the 1940s. More stringent record keeping began in the 1940s and illustrates that hatchery fry production has varied over the past half century (Table 3.1). Fry production was greatest in the 1940s and 1950s and lowest in the 1980s. In fact, no hatchery production occurred from 1983 to 1989. Increases in fry stocking after 1991 reflect the ramp-up of pike production at the Waterville hatchery.

Winter Rescue

Beginning in the late 1940s and building through the 1970s, large, shallow winterkill lakes were used as sources of fingerling, yearling, and even adult northern pike for stocking. Winterkill lakes have dissolved oxygen levels that drop low enough during severe winters to kill resident fish. Mature pike from connected waters moved into the natural spawning and nursery habitat provided by winterkill lakes, and the young pike they produced tended to remain in these areas. Trapping techniques were developed to remove (or "rescue") the bulk of pike and stock them elsewhere (Figures 3.11–3.13). During the main era of winter rescue, the technique was considered part of a multipurpose management

TABLE 3.1.
Statewide stocking records for northern pike fry, fingerlings, and yearlings plus adults. Fry were produced from hatcheries; fingerlings were from managed spawning areas and winter rescue; and yearlings plus adults were from winter rescue.

Year	Fry Number of lakes	Fry Number of fish	Fingerlings Number of lakes	Fingerlings Number of fish	Yearlings + Adults Number of lakes	Yearlings + Adults Number of fish
1936	1	169,200				
1937	1	1,060,800				
1940	5	1,776,200				
1941	2	515,612				
1942	4	1,000,000	1	150	1	27
1943	2	636,000				
1944	2	2,585,632				
1945	2	928,690				
1946	3	180,000	2	120,000		
1947	3	610,000	1	80	8	3,771
1948	1	240,000	2	265	10	1,552
1949	4	1,200,000	1	15	7	60,656
1950	18	2,582,000	14	962,460	35	10,065
1951	22	2,593,670	6	187,100	64	35,572
1952	8	1,790,000	8	60,035	72	51,599
1953	12	2,195,000	3	39,442	67	21,955
1954	4	843,000	13	24,138	84	32,934
1955	11	1,279,412	21	494,229	79	51,108
1956			29	162,369	126	141,015
1957			45	403,343	112	139,657
1958			35	331,668	74	138,168
1959	1	47,000	31	290,076	104	103,223
1960	2	191,000	74	675,921	133	88,871
1961	4	959,900	151	809,082	228	182,644
1962	3	53,001	212	4,437,181	345	229,349
1963	1	400,000	235	2,553,885	316	534,945
1964	13	1,621,607	225	952,025	318	219,194
1965	5	890,000	280	1,696,509	252	84,575
1966	1	110,000	239	4,113,869	261	156,268
1967			201	1,100,878	357	253,448
1968			161	842,673	245	132,106
1969	3	282,692	168	2,743,023	244	145,719
1970	1	234,000	229	2,722,917	391	273,381
1971	2	163,700	210	2,216,735	387	212,486
1972	3	55,000	167	3,735,864	337	263,381
1973			171	3,119,479	319	175,441
1974			177	2,233,031	258	157,700
1975	5	422,800	199	3,691,133	260	134,267
1976	4	485,000	149	2,037,862	215	175,533
1977	7	1,158,392	77	906,502	220	129,804
1978	2	152,252	224	4,511,360	222	195,316
1979	1	3,055,000	151	3,374,873	226	156,158
1980	2	134,400	136	1,435,879	196	97,104
1981	1	54,729	46	560,613	141	54,804
1982	2	44,900	94	1,803,082	135	76,049
1983			92	1,601,844	134	92,112
1984			51	2,279,118	125	100,264
1985			60	1,412,661	100	53,776
1986			73	3,385,548	113	50,561
1987			30	2,068,356	56	20,972
1988			38	634,275	59	14,394
1989			22	1,017,961	28	12,187
1990	2	106,088	32	452,954	31	41,259
1991	4	563,401	43	596,976	59	23,963
1992	1	14,880	43	418,268	76	13,552
1993	8	932,012	43	634,341	42	11,984
1994	6	475,144	23	485,762	66	50,678
1995	5	383,238	9	140,776	51	11,186
1996	7	470,801	14	233,442	24	3,761
1997	18	1,105,172	10	90,742	37	6,001
1998	22	968,328	10	93,103	35	6,236
1999	20	1,115,676	8	192,789	34	15,181
2000	20	1,248,192	2	26,300	26	5,329
2001	18	2,202,185	17	53,578	31	5,748
2002	15	1,956,962	3	17,253	35	6,956
2003	18	1,063,884	7	7,903	27	5,615
2004	9	781,305	14	35,612	20	1,231
2005	16	1,070,124	26	58,320	30	3,711

program for fish, waterfowl, and wild rice production (Johnson and Moyle 1969). Today, a winter rescue program is managed only in Rice Lake, Aitkin County; the fish are stocked into other lakes to reestablish pike populations following severe winterkills or chemical reclamations, or to maintain put-and-take fisheries in urban lakes.

The uniquely efficient trapping techniques for winter rescue were described in detail by Hanson (1958) and Johnson and Moyle (1969). Early observations showed that low dissolved oxygen concentrations in winterkill lakes caused fish to move into outlet streams or into spring seepage areas providing open water and higher oxygen levels. Similar conditions were mimicked by using water-pumping stations set up in two small artificial channels excavated into the shoreline. Water was pumped from the lake through one channel, then returned through the other channel after being aerated by discharging into a large wooden box (Figure 3.14). Fish were attracted to traps placed in each channel by the current (water flowing to the pump) or by higher oxygen concentrations (water discharged from the pump). Initial wintertime

FIGURE 3.11. Winter rescue northern pike trap.
FILE PHOTOGRAPH FROM MINNESOTA DEPARTMENT OF NATURAL RESOURCES.

FIGURE 3.12. A box trap for northern pike winter rescue excavated into the shoreline.
FILE PHOTOGRAPH FROM MINNESOTA DEPARTMENT OF NATURAL RESOURCES.

FIGURE 3.13. Loading winter rescue northern pike into a transport truck.
FILE PHOTOGRAPH FROM MINNESOTA DEPARTMENT OF NATURAL RESOURCES.

FIGURE 3.14. Water-pumping station for winter rescue.
FILE PHOTOGRAPH FROM MINNESOTA DEPARTMENT OF NATURAL RESOURCES.

catches tended to be greater in the inlet channel, which simulated water current outflowing from the lake. However, when dissolved oxygen levels dropped, the fish became more attracted to outlet water from the pump that had higher oxygen concentrations. The difference in oxygen concentrations between discharge water and the lake was often as much as 5 to 7 milligrams per liter (Johnson and Moyle 1969), a difference that served as an effective attractant.

An annual cycle of fish removal from the winterkill lakes could also consist of trap netting in open water during late autumn, then setting box (screen) traps under the ice with long leads to intercept northern pike movement. The box traps were also fished at spring holes and lake outlets after freeze-up but before critical oxygen levels of about 2 milligrams per liter were reached. When oxygen concentrations dropped below 2 milligrams per liter, the pumping stations were activated. Alternately starting and stopping

FIGURE 3.15. Distributing aerated water over a box trap.
FILE PHOTOGRAPH FROM MINNESOTA DEPARTMENT OF NATURAL RESOURCES.

the pumps at 24-hour intervals attracted more fish than continuous pumping (Johnson and Moyle 1969). Fish were removed from traps at least daily and transported in insulated tank trucks to other lakes for stocking. Winter handling and transport of the fish were done carefully to avoid harming both the fish and the people involved when air temperatures dropped substantially below 0°F (-17°C) (Figure 3.15).

Laura Lake, a 588-hectare (1,454-acre) lake near the town of Remer in Cass County, had the most intensively studied winter rescue operation. Annual catches of northern pike from Laura Lake ranged from 6.2 to 32.7 kilograms per hectare (5.5 to 29.2 pounds per acre) during the 13 winters from 1955 through 1968 (Johnson and Moyle 1969). Fluctuations in yields each year were attributed to annual differences in the severity of winterkill and their effect on the age structure of the pike population the following year. After a year of severe winterkill, when there was little carryover of fingerlings, the next winter's catch was mainly new fingerlings and a few adults remaining from the spawning run. The new fingerlings typically had a weight of 4 to 6 fish per pound, and yields averaged 12.9 pounds per acre (range of 6.2 to 19.7 pounds per acre). Much greater yields were obtained following partial winterkills (and incomplete harvests) that occurred during some winters. Fingerlings that carried over to the next winter grew substantially and were harvested as yearlings at 2 to 5 times the fingerling weight, with individuals sometimes exceeding 1 pound. Yields averaged 22.2 pounds per acre (range of 10.9 to 29.2 pounds per acre) in years when yearlings made up most of the catch. When yearlings survived over winter in large numbers, they apparently suppressed survival of fingerlings during the next season.

The winter rescue program yielded large numbers of naturally reared northern pike for stocking. Two goals attributed to the stocking program during the heydays of winter rescue were to increase the recreational harvest of pike, and to control what were considered excessive numbers of yellow perch and other unwanted fish. And, in fact, the stocked pike had considerable impacts on fish populations in the lakes receiving them. The first study of the effects of stocking winter rescue pike in Minnesota was by Wesloh and Olson (1962). Pike were stocked into Grace Lake, a 359-hectare (885-acre) lake in Hubbard County that supported a percid community with a relatively dense population of yellow perch. A high rate of return was seen from the stocked pike in Grace Lake. Within 2 years (after stocking yearlings during the winter of 1958–59), the recreational fishery had harvested 44% of the original number of pike stocked, and 160% of the weight of fish stocked. Results such as those created enthusiasm for pike management in Minnesota over the next two decades that may have led to excessive stocking of pike in some cases.

At the peak of winter rescue stocking, numbers of yearling and adult northern pike stocked into other lakes averaged 203,000

FIGURE 3.17. Contemporary water control structure and winter rescue box structure at Rice Lake, Aitkin County.
FILE PHOTOGRAPH FROM MINNESOTA DEPARTMENT OF NATURAL RESOURCES.

munities and about recruiting too many small northern pike have drastically curtailed winter rescue stocking in pike management. At the remaining winter rescue lake, Rice Lake, Aitkin County, under-ice trapping and pumping stations are no longer operated. Rather, fish are trapped at a concrete inlet/outlet water control structure for the lake, where outflow created through box traps at the structure (Figure 3.17) attracts the pike. From 2000 through 2005, numbers of yearling and adult pike taken from Rice Lake averaged only about 5,000 fish per year.

Managed Spawning and Rearing Areas

Intensive development of lakeshore property and draining of spawning sloughs prior to the early 1950s (habitat alteration that

FIGURE 3.18. (a) Water-pumping station and (b) outlet control structure for manipulating water levels during spring in Cedar Pond, Rice County.

was particularly evident in agricultural areas and the Twin Cities metropolitan area) helped motivate the MNDNR to actively develop spawning areas for northern pike as a management technique. Spawning areas included both natural sloughs and constructed ponds that were either owned by the state or were privately owned and operated in cooperation with landowners and sportsmen's clubs. Beginning in 1953, water control devices were installed in some sloughs to maintain constant water levels under the assumption that stable water was important for pike production (Adelman 1969). Control dams, dikes, channels, and in some cases water-pumping stations were placed at inlets to the spawning sloughs (Figure 3.18). Prior to spawning, slough water levels were raised by closing the dam to capture spring runoff or, if necessary, by pumping water into the sloughs.

With enough runoff water, spawning fish naturally moved into the sloughs. Otherwise, adult northern pike were stocked into the sloughs to allow natural spawning (Figure 3.19), and high water levels were maintained after spawning to enhance development of

FIGURE 3.19. Stocking adult northern pike into a managed spawning area.
FILE PHOTOGRAPH FROM MINNESOTA DEPARTMENT OF NATURAL RESOURCES.

FIGURE 3.20. Managed rearing areas stocked with northern pike fry in early spring 2003. (a) Cedar Pond, Rice County; (b) Gorman Pond, Le Sueur County; and (c) Duck Pond, Blue Earth County.

juvenile pike up to lengths of 50 to 150 millimeters (2 to 6 inches). Yet another alternative has been stocking newly hatched fry from hatcheries into managed rearing areas, then allowing them to grow to larger sizes for stocking. Fingerling pike derived from all these methods were released back into the lake or were trapped and stocked into other lakes when the sloughs were drained. Jarvenpa (1962) provided some guidelines for basin size and shape, water control, and types of vegetation that provide conditions for good pike production. The guidelines recommended shallow, completely drainable basins of 8 hectares (20 acres) or less that provided grasses and sedges for habitat (Figure 3.20).

Several intensive studies of hatching success and survival of northern pike fry in managed spawning areas were carried out from 1955 to 1967 (Franklin and Smith 1963; Bryan 1967; Adelman 1969). Many of the important findings from those studies were already reviewed in the earlier section "Reproductive Ecology" (see chapter 1). Nevertheless, it is important to highlight that water quality, food resources available for the fry, and cannibalism had important effects on hatching success, survival, and growth of the fry. Maximum pike production depended on control of water levels, manipulation of water flow, and maintaining appropriate types of vegetation in the wetlands.

The contribution that fish from a managed spawning area made to a natural year class (produced in-lake) was estimated in Reeds Lake, a 76-hectare (187-acre) water in Waseca County (Woods 1963). Natural reproduction within Reeds Lake was augmented in 1958 with an adjacent 1.2-hectare (3-acre) managed spawning marsh. The lake level was low in 1958, so much of the natural in-lake spawning habitat was unavailable to northern pike. About as many pike were produced from the managed spawning area as from all the natural areas combined, but the estimated combined production of 5,000 fingerlings resulted in a weak year class compared with other years (Woods 1963). In Lake George, a 185-hectare (456-acre) Anoka County lake, fingerling production

Size selectivity of gill nets, essentially the sizes of pike most readily caught in the meshes, has been described by Frost and Kipling (1967), Neumann and Willis (1994), and Pierce et al. (1994). Size selectivity can be estimated indirectly by graphically comparing numbers of fish from various size classes caught in different mesh sizes (McCombie and Frie 1960). Graphical information for all size classes of fish can then be incorporated into a master curve (Figure 4.4) that shows the efficiency of gill nets in relation to ratios of the length of the fish to the perimeter of the mesh size where they were caught. Gill nets are most effective for catching pike when the ratio of the length of the fish to the perimeter of the mesh size is 3.2 to 4.5 (Figure 4.4) (Pierce et al. 1994).

Finally, the master curve can be used to project relative efficiencies of the whole gang of five meshes to catch pike of various sizes. Figure 4.5 shows the effectiveness of our multimesh experimental nets for different sizes of pike after adjusting for the rate at which different sizes of pike might encounter the nets (differing swimming performance with pike size) (Pierce et al. 1994).

Lengths of northern pike observed during lake surveys in Minnesota over the past half century have been a reflection of both gill-net selectivity and existing fish populations. The experimental gill nets are not efficient for catching young-of-the-year and yearling pike, in part because lake survey nets are not usually fished in the very shallow, vegetated water preferred by small pike. Relative efficiency also seems to drop below 60% for fish larger than 800 millimeters (31.5 inches) total length (Figure 4.5). The dip in efficiency around 550 to 600 millimeters (21.7 to 23.6 inches) (see Figure 4.5) resulted from the 13-millimeters increase in mesh size between the largest two meshes in experimental gill nets compared with 6-millimeters increases between smaller mesh sizes. An interesting observation is that the statewide average length for pike sampled in experimental gill nets has been 507 millimeters (20 inches) (n = 137,838 fish from 4,196 lake surveys from 1950 through 1990), which corresponds to the peak of the gill-net selectivity curve (Figure 4.5) (Pierce et al. 1994). Use of five mesh sizes, however, has made the overall selectivity curve for experimental nets broad enough to help discern differences in size structure among pike populations, as well as historical changes in individ-

TABLE 4.1.
Population sizes (estimated from tagging studies of northern pike) in Lake of Isles, Itasca County, compared to projected numbers of pike for gill-net catches.

Fish length (millimeters)	Population estimates		Corrected gill-net catches	
	Number	Proportion	Number	Proportion
351–400	218	0.25	85	0.26
401–450	312	0.35	108	0.33
451–500	238	0.27	87	0.27
501–550	108	0.12	38	0.12
551–600	10	0.01	5	0.02

Note: Proportions listed are proportions of the pike population between 351 and 600 millimeters total length. Corrected gill-net catches are projected from the actual numbers of pike caught in gill nets divided by relative efficiency of the gear for each length class (Table 4.2).

TABLE 4.2.
Relative capture efficiencies estimated for various length classes (presented in both millimeters and inches) for northern pike in multimesh experimental gill nets.

Fish length (millimeters)	Relative efficiency	Fish length (inches)	Relative efficiency
251–300	0.54	10–12	0.53
301–350	0.74	12–14	0.76
351–400	0.82	14–16	0.81
401–450	0.94	16–18	0.92
451–500	1.00	18–20	1.00
501–550	1.00	20–22	0.95
551–600	0.84	22–24	0.78
601–650	0.79	24–26	0.79
651–700	0.81	26–28	0.77
701–750	0.72	28–30	0.68
751–800	0.61	30–32	0.57
801–850	0.50	32–34	0.47
851–900	0.41	34–36	0.37

ual populations. For example, among 44 ecological lake types that differ physically and chemically (Schupp 1992), average lengths of gill-netted northern pike have ranged from 455 to 587 millimeters (17.9 to 23.1 inches).

Results of size selectivity studies provide a useful framework for interpreting gill-net catches. Relative efficiencies of the nets for different size classes of northern pike can be used to predict what the actual population size structures look like. A corrected size structure for pike can be provided by dividing raw numbers of fish caught in gill nets by relative efficiencies for various size classes of pike, as was done for Lake of Isles, a 25-hectare (62-acre) Itasca County lake (Table 4.1) (Pierce et al. 1994). The corrected size structure in Lake of Isles was then compared with population sizes estimated from tagging studies. Proportions of fish in 50-millimeter length classes calculated from corrected gill-net catch data were nearly identical to the length structure from population estimates (Table 4.1). The accuracy was due, in part, to the extensive amount of gill netting done on Lake of Isles and to the limited size range of fish encountered. Relative efficiencies for correcting net catch data (for fish lengths in both millimeters and inches) are provided in Table 4.2.

Comparing Gill-Net Catch Rates

Catch rates from gill nets (i.e., numbers of northern pike per overnight experimental net set) can be used as indices to compare relative abundance of pike among lakes, and changes in abundance over time within individual lakes (Figure 4.6). Such direct comparisons need to consider the attributes of catch-rate data, which can vary greatly from net to net and can have numerous zero values. In view of these statistical attributes, Moyle and Lound (1960) recommended comparing catch rates using (1) a parametric approach treating the data as a negative binomial distribution, or (2) a nonparametric data-ranking approach that essentially compares median catch rate values. Although medians appear most useful for comparing catch rates, they have the disadvantage of being zero if more than half the nets catch no pike. These approaches mean little to the layperson who does not have a background in statistical theory, but they are nevertheless important considerations for comparing catch rates.

Investigators in other geographic areas have found gill-net catch rates for northern pike to decline between early and late summer (Casselman 1978b; Diana 1980; Cook and Bergersen 1988). A large-scale analysis of gill-net catch rates from 433 Minnesota lakes from 1983 through 1997 also illustrated a decline in pike catch rates through the summer months (Grant et al. 2004); however, the decline did not represent a very large change in catch rate between 15 June and 24 August. It is quite possible that the observed decline in catch rates was actually reflecting mortality of pike during the summer because total mortality estimates for pike often exceed 50% annually. Movement of some fish into cooler water not sampled by gill nets is another potential mechanism for the low-level seasonal decline in catch rate. Regardless, repeat sampling from the same lake on similar dates provides the most valid historical comparisons.

Directly comparing catch rates also makes the important assumption that gill nets fished in different lakes or during different years in the same lake have a consistent catchability coefficient (q) where the simplest form of the relationship is

catch rate = q (fish density).

The catchability coefficient (q) is a value that relates how susceptible, or "catchable," the fish are to the sampling gear. Unfortunately, literature for other types of fisheries has shown that q can vary with sizes of the fish (because of size selectivity of the gear), with fish density (density-dependent q), or with where the fishing effort occurs compared with where the fish are actually located (Hilborn and Walters 1992). In Minnesota, we tested the ability of gill-net catches to monitor changes in northern pike

FIGURE 4.6. Northern pike caught in an experimental gill net.

population numbers by removing fish from 28-hectare (70-acre) Camerton Lake near Grand Rapids in Itasca County (Pierce and Tomcko 2003b). Results showed that gill-net catch rates tracked the declining population size of pike relatively well until the removals started to cause shifts in the ages, sizes, and possibly

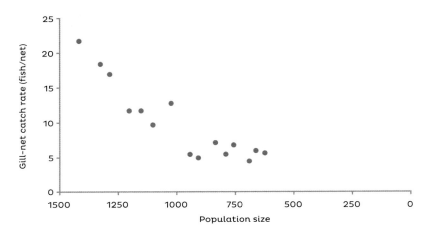

FIGURE 4.7. Northern pike population size (number of fish age 2 and older) compared with gill-net catch rates as pike were removed from Camerton Lake, Itasca County.

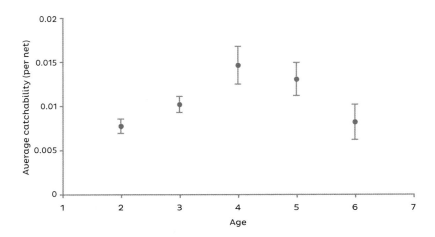

FIGURE 4.8. Average catchabilities for northern pike ages 2 through 6 caught in gill nets from Camerton Lake, Itasca County. Vertical bars represent potential statistical error. Average lengths at capture were 412, 458, 519, 551, and 582 millimeters for ages 2 through 6.

ages of fish sacrificed. The percentage of the population sacrificed was negatively related to lake size, with the biggest increases in percentages sacrificed found in the smallest lakes. The significance of this nonlinear relationship was evident when proportions of the population killed were plotted against the lake sizes (Figure 4.10). Among the 15 lakes, average percentages killed were 1.1% for lakes greater than 600 acres (sample size of 3 lakes), 2.2% for 300- to 599-acre lakes (4 lakes), 3.7% for 100- to 299-acre lakes (5 lakes), and 7.0% for lakes less than 100 acres (3 lakes). Mortality rates resulting from fish sacrificed during survey netting are comparable to recreational exploitation rates in some cases and should therefore be considered carefully in developing long-term monitoring strategies that involve gill netting. Frequent gill netting could alter pike population dynamics.

Spring Trap Netting

Spring trap netting in north-central Minnesota lakes has produced large numbers of northern pike as soon as a small margin of open water develops along shorelines, inlets, outlets, and shallow bays where ice first melts away (Figure 4.11). Trapping as the ice recedes takes advantage of the tendency of pike to move into the warmest water they can find early in the spring, when they are also starting to stage for spawning. Catch rates in the trap nets remain fairly high until a few days after ice-out, when catch rates along shorelines decline dramatically. Analysis of spring-trapping dates over a 14-year period in north-central Minnesota (using a total of 26 lake-years of information) showed that on average the most productive spring trapping was during the 11 days from 14 April to 25 April. The most productive trapping period lasted 5 to 18 days depending on spring weather patterns and how the shape and orientation of the lake basin influenced melting of the ice.

During the short time period while ice is receding, large proportions of the northern pike population can be sampled in a lake.

The average proportion of the population of fish age 2 and older sampled by spring trap netting was 29% across 15 north-central Minnesota lakes ranging in size from 15 to 765 hectares (38 to 1,890 acres). In 7 of the lakes, we sampled over 33% of the pike population. Spring trapping can be extremely efficient where spring fish movements and spawning sites are confined to a small area that can be easily trapped. For example, 92% of the estimated population of fish age 2 and older was trapped when fish movements in spring were concentrated along one side of a bay in the southern end of Willow Lake, Itasca County. Spring trapping can also be very effective in small lakes. Fifty-four percent of the pike population was trapped in 15-hectare Forest Lake, Itasca County. In contrast, pike populations can be much harder to trap in larger lakes where spring movements are more dispersed. Such was the case in Coon-Sandwick Lake, Itasca County, where extensive trapping sampled only 5% of the estimated population of fish age 2 and older.

FIGURE 4.11. Spring trap netting along the edge of the ice.

Just about any type of trap net can be used as a nonlethal gear for catching northern pike during early spring movements, including the extra-large framed nets used for muskellunge assessments. Standard lake survey trap nets in Minnesota have two rectangular 3 × 6 foot (0.9 × 1.8 meter) metal frames, six steel or fiberglass hoops with two throats, a 40-foot (12.2-meter) "lead," and 0.75-inch (19-millimeter) bar measure mesh size, whereas muskellunge nets have a 5 × 6 foot (1.5 × 1.8 meter) frame. The lead is a fine-mesh net that extends from the trap into shore (Figure 4.12). As fish swim along the shoreline, they encounter the lead and follow it into deeper water and into the throats of the trap net. Trap nets using heavy twine for the outside edges of the rectangular throat in the frame can catch fish of larger girths because the throat is more elastic than frames using metal rods as the outside edges. The MNDNR has often used 1-inch (25-millimeter) bar mesh nets for spring trapping to avoid incidental catch of smaller fish species.

FIGURE 4.12. Trap net being set from a boat with the lead of the trap net extending into shore.

Clark and Willis (1989) found no significant difference in pike sizes or trap-net catch rates of pike captured in 0.5-inch (13-millimeter) versus 1-inch (25-millimeter) mesh sizes. Davis and Schupp (1987) found that the average lengths and catch rates of pike in modified single-frame trap nets were not different than in nets with two frames.

Mortality rates for fish that are trapped seem to be very low during spring when water temperatures are cold. In Trout Lake near Coleraine, Itasca County, 1,453 northern pike larger than 300 millimeters (11.8 inches) total length were caught by trap netting. Of these, 2 fish (0.1%) were dead, and 9 were in somewhat rough shape. Four more small fish caught around the gills were found in the leads when the nets were removed. If all these fish eventually died, trapping mortality was still only 1.0% (95% confidence limits = 0.6% to 1.7%). The low mortality of trapped pike in Trout Lake was also evident from the large numbers of those fish that were recaptured. Over a 16-day trapping period, 378 fish were recaptured at least once, 106 fish were recaptured at least twice, 20 fish were recaptured three times, and 2 fish were recaptured four times. Mortality of pike that were trapped, tagged, and then held in net pens for 5 days was determined in Lake of Isles, Itasca County. The result of handling the fish plus holding them in net pens was a mortality rate of 2.4% (Pierce and Tomcko 1993).

Short-Term Spring Gill Netting

Short-term (3 to 4 hours) gill netting during the middle of the day in spring while water temperatures are still cool (less than about 15°C [60°F]) offers an alternative method for nonlethal sampling. The combination of cool water temperatures and short duration of the net sets makes mortality and stress on the fish about equivalent to catch-and-release angling even though standard lake survey gill nets are used. For example, mortality of northern pike caught using short-term sets in Camerton Lake, Itasca County, during spring 1998 was 6.2% of 193 fish. Water temperatures during the netting were 51° to 58°F (10.7° to 14.7°C), and the average set time was 4.0 hours (range = 3.3 to 5.0 hours). Moreover, survival of fish tagged from those short-term gill net sets was at least as good as survival of fish tagged during spring trap netting in 1998. Intensive netting from mid-June through July in Camerton Lake recovered 50.8% of fish tagged from spring gill netting (60 of 118 tagged fish), whereas 40.4% of fish tagged from trap nets were recovered (269 of 666 tagged fish). By comparison, nearly all pike die during summer lake survey gill netting when water temperatures are warmer and the nets are fished overnight.

In addition to comparatively low mortality, short-term gill netting during spring offers other useful advantages. First, net locations in each lake can be strictly randomized. Randomization is a key approach for meeting assumptions of mark-recapture experiments and for assuring unbiased sampling of fish populations. Second, the midday net sets often have very low incidental catches of other fish species, reducing impacts to other fish populations and minimizing the effort required to pick all of the fish out of the nets. An exception occurs where small yellow perch are abundant and get caught in large numbers in the smallest mesh size. Finally, work crews can avoid hazardous weather that sometimes occurs during spring. Because the nets are set and lifted on the same day, work crews can choose to avoid fishing the nets during exceptionally hazardous days.

Comparisons of Netting Techniques

Catches from spring trap netting, spring short-term gill netting, and summer (lake survey) gill netting can be contrasted in lakes where all three methods were used in the same year. Catches from the three sampling methods were compared in 14 lakes where the density of northern pike greater than 350 millimeters (13.8 inches) total length was also estimated (Table 4.4). Spring sampling encountered large numbers of northern pike, but comparisons of catch rates and population density estimates among the 14 lakes showed that catch rates during the spring were not good indices of pike abundance compared with summer gill netting. No relationships were found between pike densities and their catch rates during either spring trap netting or short-term gill netting. Nor were the highest daily trap-net catch rates each spring (peak catch rates) related to pike densities. Weather-related movement patterns of pike apparently had large influences on their catch rates during the spring. Summer gill-net catches, on the other hand, were more closely related to pike densities.

TABLE 4.4.
Population density estimates, peak catch rates for spring trap netting, and average lengths and catch rates for spring trap netting, short-term gill netting in spring, and summer lake survey gill netting in 14 north-central Minnesota lakes.

Lake	Average length (millimeters)			Average catch rate (fish/net)			Spring trap net peak catch rate (fish per net)	Population density (fish per hectare)
	Spring trap net	Spring gill net	Summer gill net	Spring trap net	Spring gill net	Summer gill net		
Camerton	460	477	473	9.94	8.77	21.67	14.08	59.0
Chase	586	577	582	3.79	3.25	6.38	11.00	8.0
Coon-Sandwick	477	454	456	2.44	3.92	8.00	2.93	32.4
Forest	557	596	632	2.74	5.77	1.75	2.88	9.2
Medicine	457	530	532	7.19	4.36	9.50	18.00	36.5
North Twin	484	522	565	1.79	5.07	9.00	4.33	13.8
Ruby	492	515	494	5.53	3.88	7.50	10.57	14.0
Sand	465	488	463	4.50	7.45	6.00	6.93	26.3
Sissabagamah	457	454	451	4.62	5.93	5.22	11.36	24.0
Six Mile	483	548	526	9.37	2.72	2.92	17.64	8.6
Snaptail	507	492	516	1.98	5.11	6.83	3.27	19.9
Trout	590	612	628	4.97	4.65	4.67	8.12	4.9
Wilkins	501	502	481	5.61	4.05	6.33	14.60	11.9
Willow	516	561	560	3.55	7.94	2.00	5.29	3.2

All three netting techniques caught a wide range of northern pike sizes, although spring trap netting caught a slightly greater range of fish sizes (158 to 1,150 millimeters [6.2 to 45.3 inches] total length) than gill nets (211 to 986 millimeters [8.3 to 38.8 inches]). Paired comparisons of average lengths of fish caught by the different gear types in each lake (Table 4.4) showed no differences between average lengths of fish caught in spring short-term gill netting and summer gill netting. The average lengths of pike caught during spring trapping were often lower than either method of gill netting, but in three lakes trap-net average lengths were actually greater (Table 4.4; also see Figure 4.13).

FIGURE 4.13. Three large northern pike caught by a Minnesota Department of Natural Resources crew during spring trapping.
PHOTOGRAPH COURTESY OF TONY KENNEDY.

Seven of the lakes in Table 4.4 had estimated population numbers for different size categories of northern pike. In those lakes, average lengths were projected from the population estimates (based on sums of products of the population estimates and the midpoints of each size range). In theory, this method should approximate the true average length for the pike populations and provide an interesting comparison with average lengths observed from each sampling method. In reality, however, these projected average lengths were probably not completely independent of gear biases, since both trap netting and gill netting were used to derive the population estimates (see the section below titled "Population Estimation"). Projected average lengths were as follows: Forest Lake = 563 millimeters (22.2 inches), Ruby Lake = 493 millimeters (19.4 inches), Sand Lake = 447 millimeters (17.6 inches), Snaptail Lake = 493 millimeters (19.4 inches), Trout Lake = 597 millimeters (23.5 inches), Wilkins Lake = 478 millimeters (18.8 inches), and Willow Lake = 533 millimeters (21.0 inches). These length projections were similar to average lengths from spring trapping, but average lengths for spring and summer gill netting were typically greater. The results imply that spring trapping provides the most accurate picture of lengths for a pike population.

Sampling Larval Northern Pike

Nursery habitat for very young northern pike is difficult to sample because it consists primarily of marshes or shallow water with submerged and emergent vegetation. Conventional methods for sampling larval fish in open water, such as ichthyoplankton (very fine mesh) nets towed from a boat, are not very useful for sampling these kinds of habitats. Small-mesh bag seines (sometimes lined with cheesecloth) are usually dragged through the water by two people in waders and have been used to sample juvenile northern pike (fish larger than about 35 millimeters [1.4 inches] that resemble the adult form). Seining is less efficient in submerged vegetation than in other types of habitat, so northern pike relative abundance is underestimated in submerged vegetation (Holland and Huston 1984). Intensive studies of the early life history stages of northern pike in Minnesota have also used wire-mesh scap nets to sample small pike (Adelman 1969). Scap nets are stiff, fine mesh, triangular-shaped nets on a long handle that are dipped through the water. Studies in New York sampled nursery areas using an enclosure swept with a dip net to remove fish (Forney 1968; Morrow et al. 1997), but it is not clear how efficient the method is for catching different sizes of northern pike.

Lighted Plexiglas "Quatrefoil" traps have been an effective method of collecting larval and juvenile northern pike (Zigler and Dewey 1995; Pierce et al. 2006). Young pike are positively phototactic, so they are drawn into the traps by artificial light. Quatrefoil traps consist of four Plexiglas cylinders, open to each other in the center, with a central light system (Figures 4.14-4.15). Light traps offer the important advantage of minimally disrupting nursery

FIGURE 4.14. Quatrefoil light trap for sampling larval and juvenile northern pike. This light trap features 6-millimeter entrance slots between four Plexiglas cylinders and a light-emitting diode (LED) that illuminates a fiber-optic strip wrapped in a spiral around a central rod.

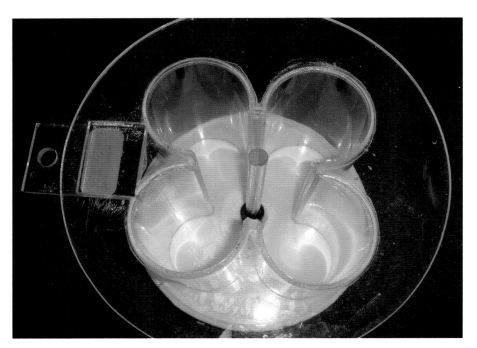

FIGURE 4.15. Bottom view of a Quatrefoil light trap with the metal collection bowl removed, illustrating the cloverleaf shape of the inside of the light trap.

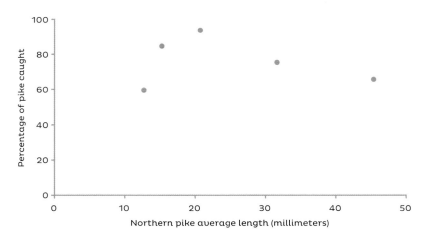

FIGURE 4.16. Percentages of northern pike, at various lengths, caught in Quatrefoil light traps. Pike were stocked into 5-meter (16-foot) hatchery raceways with a light trap located near each end. The light traps were fished for 2 hours after hatchery lights were turned off at night.

habitat compared with other sampling techniques. In our work with light traps, we (Pierce et al. 2006) demonstrated that light-trap catch rates discriminated between different densities of larval northern pike stocked into hatchery raceways. Light traps effectively caught all sizes of pike ranging from the stage when larvae first became active (12 to 13 millimeters [about 0.5 inches] total length) (Figure 4.16) until they became too large to fit through the trap entrance (>66 millimeters [2.6 inches] total length). In fact, the raceway trials showed that light trapping could be a very efficient method for sampling larval pike. In one set of trials (the highest point on Figure 4.16), light traps caught 57 of 60 fish planted in one raceway, and 55 of 60 fish planted in the other raceway during a 2-hour period. Furthermore, light traps were capable of detecting patchy fish distributions as well as illustrating temporal changes in the numbers and sizes of juvenile pike in managed wetlands (Pierce et al. 2006). Projected uses of light trapping in managed nursery areas include monitoring mortality and growth and identifying the most productive wetlands, stocking rates, and cover types to use for rearing pike fingerlings. Light traps have utility also for sampling natural nursery habitat (Pierce et al. 2007).

Tagging Northern Pike

Marking fish with unique identifiers, such as individually numbered tags, has been a useful tool for discovering northern pike migration between water bodies and for mark-recapture experiments designed to estimate population numbers and exploitation rates. In Minnesota, plastic anchor tags have been used extensively to mark pike. One method has been to double-mark: an individually numbered Floy tag (Floy FD-68B, 25-millimeter monofilament) is inserted on one side of the fish, and a batch (similar tags are used for a large number of fish) Dennison tag (Dennison 08966, 75 millimeters) is inserted on the other side (Figure 4.17). Both types of tag have T-shaped ends that anchor cross-ways in

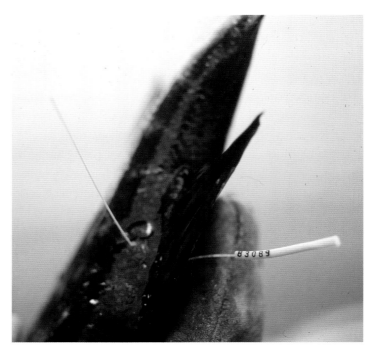

FIGURE 4.17. A Dennison tag (thin orange strand) and an individually numbered Floy tag (yellow) inserted on opposite sides of the fish near the front of the dorsal fin. Paddle-shaped ends were trimmed from the Dennison tag to leave a single thin strand of plastic about 65 millimeters long.

the fish's flesh to hold the tag in place. Tags are inserted diagonally forward and laterally using a tagging gun so that the T-shaped end of the tag anchors itself among the dorsal pterygiophores near the front of the dorsal fin. Pterygiophores are internal skeletal supports extending into the body from fin rays. Marking is quickly carried out, and no anesthetic is needed for the fish.

Fish that lose tags and mortalities caused by handling and tagging fish can cause serious biases that underestimate exploitation rates and inflate population estimates. These potential sources of bias seem to be minimal for plastic anchor tags. Annual rates of tag loss were found to be 1.8% for Floy tags and 0% for Dennison tags in Coon-Sandwick Lake, Itasca County, and the probability of a fish

A novel approach for identifying northern pike spawning habitat is implanting miniature radio transmitters through the oviduct into the egg masses of mature females just prior to spawning (Figure 4.18). The transmitters are expelled with the eggs during spawning, and spawning sites can then be found by locating the shed transmitters. A preliminary study by Pierce (2004) found that advances in miniaturization have led to transmitters that are small enough to fit up through the oviducts of pike. Insertion of miniature transmitters into the oviduct, a nonsurgical procedure, has cost- and time-saving advantages over other approaches for finding spawning habitat, and also avoids stresses on the fish caused by transmitter insertion methods that require surgery.

Additional studies (Pierce et al. 2005; Pierce et al. 2007) in Willow Lake, Itasca County, showed that northern pike deposited many of the transmitters in a previously known spawning area, the west side of a bay at the south end of the lake (Figure 4.19). Size of the transmitter in relation to fish size was important. The most consistent success was for pike larger than 700 millimeters (27.6 inches) total length; 8 of 10 large fish expelled transmitters in the known spawning area. One of the remaining fish retained its transmitter, whereas the other fish expelled its transmitter in less likely habitat. The studies in Willow Lake indicate that oviduct implantation of miniature radio transmitters is a quick and easy method of attaching radio tags to fish. Eventually, technological improvements leading to smaller batteries and transmitters should allow miniature transmitters to be more consistently expelled with the eggs or perhaps even to be implanted within eggs.

The utility of oviduct insertion of radio transmitters as a resource management technique was demonstrated in Moose Lake, Itasca County. Telemetry with oviduct-implanted transmitters showed that both northern pike and a coexisting muskellunge population used various habitats in Moose Lake during spring

FIGURE 4.19. The nineteen locations of transmitter deposition sites for all sizes of northern pike in Willow Lake during 2002 and 2003. One cluster of points at the very southwest end of the lake contains nine triangles, but because of the close proximity of transmitters deposited in this important spawning location, several triangles are obscured by others.

FIGURE 4.20. Researcher with a hand-held loop antenna radio-tracking northern pike during early spring.

excessive regulation, and still others enjoy the tradition of darkhouse spearing. These people have very different opinions about what pike fishing should look like in the future, and any changes in pike management elicit heated responses and even backlash in the state's legislature.

Fortunately, Minnesota is blessed with a huge northern pike resource; pike populations are plentiful in lakes, streams, and rivers throughout the state. They are plentiful enough that pike management today is a matter of trying to provide adequate fishing opportunities for the different fishing groups. Therefore, pike may be managed differently from one lake to another, and the critical issue is getting public consensus on how many lakes should be managed for bigger pike with special regulations, and where those lakes should be located (Figure C.1). During summer 2010, research surveys (questionnaires) from the University of Minnesota Department of Fisheries, Wildlife and Conservation Biology were sent to licensed pike anglers and spearers. The survey's purpose is to get better feedback on how well the interests of various pike fishing groups are being balanced. Hopefully the surveys will provide some guidance about the future use of special regulations in Minnesota.

The second important question affecting the future is, to what extent will human population growth and land development alter shorelines and watersheds surrounding public waters? How can the public interests in fishing, aquatic habitats, and water quality be balanced with private landowners' interests in developing their properties? Casselman and Lewis (1996) and Margenau et al. (2008) have documented that the incremental loss of wetlands, changes in shoreline cover and structure, and addition of excess nutrients to the water have affected water quality and clarity, aquatic plants, and, ultimately, the habitat of northern pike. From a lake stewardship standpoint, it is clear that whatever we do to the land affects the water and the fish in it. Yet, time and again we see new instances of natural shorelines that are altered to create lawns,

FIGURE C.2. Filamentous algae bloom as a result of lawn runoff at an Itasca County lake. The rest of the shoreline around the lake had clear water, dramatically illustrating how one lawn can pollute the water.

Bibliography

Aadland, L. P., and A. Kuitunen. 2006. Habitat suitability criteria for stream fishes of Minnesota. Minnesota Department of Natural Resources Special Publication 162, St. Paul.

Adelman, I. R. 1969. Survival and growth of northern pike (*Esox lucius* L.) in relation to water quality. Doctoral dissertation. University of Minnesota, St. Paul.

Adelman, I. R., and L. L. Smith Jr. 1970. Effect of oxygen on growth and food conversion efficiency of northern pike. Progressive Fish-Culturist 32:93–96.

Allen, M. S., L. E. Miranda, and R. E. Brock. 1998. Implications of compensatory and additive mortality to the management of selected sportfish populations. Lakes and Reservoirs: Research and Management 3:67–79.

Anderson, D. W., and D. H. Schupp. 1986. Fish community responses to northern pike stocking in Horseshoe Lake, Minnesota. Minnesota Department of Natural Resources, Section of Fisheries Investigational Report 387, St. Paul.

Anderson, R. O., and S. J. Gutreuter. 1983. Length, weight, and associated structural indices. Pages 283–300 *in* L. A. Nielsen and D. L. Johnson, editors. Fisheries techniques. American Fisheries Society, Bethesda, Md.

Appleget, J. G. 1951. Darkhouse spearing of northern pike in Itasca County, Minnesota, during the seasons of 1949–50 and 1950–51. Presentation for the 13th Midwest Fish and Wildlife Conference, Minneapolis, Minn.

Arlinghaus, R., T. Klefoth, S. J. Cooke, A. Gingerich, and C. Suski. 2009. Physiological and behavioral consequences of catch-and-release angling on northern pike (*Esox lucius* L.). Fisheries Research 97:223–33.

Arlinghaus, R., T. Klefoth, A. Kobler, and S. J. Cooke. 2008. Size selectivity, injury, handling time, and determinants of initial hooking mortality in recreational angling for northern pike: the influence of type and size of bait. North American Journal of Fisheries Management 28:123–34.

Armstrong, J. D. 1986. Heart rate as an indicator of activity, metabolic rate, food intake and digestion in pike, *Esox lucius*. Journal of Fish Biology 29 (Supplement A):207–21.

Armstrong, J. D., M. C. Lucas, I. G. Priede, and L. De Vera. 1989. An acoustic telemetry system for monitoring the heart rate of pike, *Esox lucius* L., and other fish in their natural environment. Journal of Experimental Biology 143:549–52.

Beaudoin, C. P., W. M. Tonn, E. E. Prepas, and L. I. Wassenaar. 1999. Individual specialization and trophic adaptability of northern pike *(Esox lucius)*: an isotope and dietary analysis. Oecologia 120:386-96.

Becker, G. C. 1983. Fishes of Wisconsin. University of Wisconsin Press, Madison.

Beukema, J. J. 1970. Acquired hook-avoidance in the pike *Esox lucius* L. fished with artificial and natural baits. Journal of Fish Biology 2:155-60.

Beyerle, G. B. 1971. A study of two northern pike-bluegill populations. Transactions of the American Fisheries Society 100:69-73.

Beyerle, G. B., and J. E. Williams. 1968. Some observations of food selectivity by northern pike in aquaria. Transactions of the American Fisheries Society 97:28-31.

Borge, L. J., and J. A. Leitch. 1988. Winter darkhouse spearing in Minnesota: characteristics of participants. North Dakota State University, Tri-College University Center for Environmental Studies, Miscellaneous Publication 2, Fargo.

Bosworth, A., and J. M. Farrell. 2006. Genetic divergence among northern pike from spawning locations in the upper St. Lawrence River. North American Journal of Fisheries Management 26:676-84.

Bry, C. 1996. Role of vegetation in the life cycle of pike. Pages 45-67 *in* J. F. Craig, editor. Pike biology and exploitation. Chapman and Hall, London.

Bryan, J. E. 1967. Northern pike production in Phalen Pond, Minnesota. Journal of the Minnesota Academy of Science 34:101-9.

Bryan, M. D., and D. L. Scarnecchia. 1992. Species richness, composition, and abundance of fish larvae and juveniles inhabiting natural and developed shorelines of a glacial Iowa lake. Environmental Biology of Fishes 35:329-41.

Brynildson, C. 1958. What's happening to northern pike spawning grounds? Wisconsin Conservation Bulletin 23(5):1-3.

Buck, D. H., and C. F. Thoits. 1965. An evaluation of Petersen estimation procedures employing seines in 1-acre ponds. Journal of Wildlife Management 29:598-621.

Burns, D. C. 1991. Cumulative impacts of small modifications to habitat. Fisheries 16(1):12-14.

Buynak, G. L., and H. W. Mohr Jr. 1979. Larval development of the northern pike *(Esox lucius)* and muskellunge *(Esox masquinongy)* from northeast Pennsylvania. Proceedings of the Pennsylvania Academy of Science 53:69-73.

Carbine, W. F. 1943. Egg production of the northern pike, *Esox lucius* L., and the percentage survival of eggs and young on the spawning grounds. Papers of the Michigan Academy of Science, Arts, and Letters 29:123-37.

———. 1944. Growth potential of the northern pike *(Esox lucius)*. Papers of the Michigan Academy of Science, Arts, and Letters 30:205-20.

Carlander, H. B. 1954. History of fish and fishing in the Upper Mississippi River. Publication of the Upper Mississippi River Conservation Committee.

Carlson, B. M. 2007. Beneath the surface: a natural history of a fisherman's lake. Minnesota Historical Society Press, St. Paul.

Casselman, J. M. 1974. External sex determination of northern pike, *Esox lucius* Linnaeus. Transactions of the American Fisheries Society 103:343-47.

———. 1978a. Calcified tissue and body growth of northern pike, *Esox lucius* Linnaeus. Doctoral dissertation. University of Toronto.

———. 1978b. Effects of environmental factors on growth, survival, activity, and exploitation of northern pike. Pages 114-28 *in* R. L. Kendall, editor. Selected coolwater fishes of North America. American Fisheries Society, Special Publication 11, Bethesda, Md.

———. 1990. Growth and relative size of calcified structures of fish. Transactions of the American Fisheries Society 119:673-88.

———. 1996. Age, growth and environmental requirements of pike. Pages 69-101 *in* J. F. Craig, editor. Pike biology and exploitation. Chapman and Hall, London.

Casselman, J. M., E. J. Crossman, P. E. Ihssen, J. D. Reist, and H. E. Booke. 1986. Identification of muskellunge, northern pike, and their hybrids. Pages 14-46 *in* G. E. Hall, editor. Managing muskies: a treatise on the biology and propagation of muskellunge in North America. American Fisheries Society, Special Publication 15, Bethesda, Md.

Casselman, J. M., and C. A. Lewis. 1996. Habitat requirements of northern pike *(Esox lucius)*. Canadian Journal of Fisheries and Aquatic Sciences 53 (Supplement 1):161-74.

Chapman, L. J., and W. C. Mackay. 1984a. Direct observation of habitat utilization by northern pike. Copeia 1984:255-58.

———. 1984b. Versatility in habitat use by a top aquatic predator, *Esox lucius* L. Journal of Fish Biology 25:109-15.

———. 1990. Ecological correlates of feeding flexibility in northern pike *(Esox lucius)*. Journal of Freshwater Ecology 5:313-22.

Chapman, L. J., W. C. Mackay, and C. W. Wilkinson. 1989. Feeding flexibility in northern pike *(Esox lucius)*: fish versus invertebrate prey. Canadian Journal of Fisheries and Aquatic Sciences 46:666-69.

Clark, C. F. 1950. Observations on the spawning habits of the northern pike, *Esox lucius*, in northwestern Ohio. Copeia 1950:285-88.

Clark, S. W., and D. W. Willis. 1989. Size structure and catch rates of northern pike captured in trap nets with two different mesh sizes. Prairie Naturalist 21:157-62.

Coble, D. W. 1973. Influence of appearance of prey and satiation of predator on food selection by northern pike *(Esox lucius)*. Journal of the Fisheries Research Board of Canada 30:317-20.

Colby, P. J., P. A. Ryan, D. H. Schupp, and S. L. Serns. 1987. Interactions in north-temperate lake fish communities. Canadian Journal of Fisheries and Aquatic Sciences 44 (Supplement 2):104-28.

Coltman, D. W. 2008. Evolutionary rebound from selective harvesting. Trends in Ecology and Evolution 23:117-18.

Conover, D. O., S. A. Arnott, M. R. Walsh, and S. B. Munch. 2005. Darwinian fishery science: lessons from the Atlantic silverside *(Menidia menidia)*. Canadian Journal of Fisheries and Aquatic Sciences 62:730-37.

Cook, M. F., and E. P. Bergersen. 1988. Movements, habitat selection, and activity periods of northern pike in Eleven Mile Reservoir, Colorado. Transactions of the American Fisheries Society 117:495-502.

Cook, M. F., and J. A. Younk. 1998. A historical examination of creel surveys from Minnesota's lakes and streams. Minnesota Department of Natural Resources, Section of Fisheries Investigational Report 464, St. Paul.

Cook, M. F., J. A. Younk, and D. H. Schupp. 1997. An indexed bibliography of creel surveys, fishing license sales, and recreational surface use of lakes and rivers in Minnesota. Minnesota Department of Natural Resources, Section of Fisheries Investigational Report 455, St. Paul.

Craig, J. F. 1996. Population dynamics, predation and role in the community. Pages 201-17 *in* J. F. Craig, editor. Pike biology and exploitation. Chapman and Hall, London.

———. 2008. A short review of pike ecology. Hydrobiologia 601:5-16.

Cross, T., and M. McInerny. 1995. Influences of watershed parameters on fish populations in selected Minnesota lakes of the Central Hardwood Forest Ecoregion. Minnesota Department of Natural Resources, Section of Fisheries Investigational Report 441, St. Paul.

———. 2006. Relationships between aquatic plant cover and fish populations based on Minnesota lake survey data. Minnesota Department of Natural Resources, Section of Fisheries Investigational Report 537, St. Paul.

Cross, T. K., M. C. McInerny, and R. A. Davis. 1992. Macrophyte removal to enhance bluegill, largemouth bass and northern pike populations. Minnesota Department of Natural Resources, Section of Fisheries Investigational Report 415, St. Paul.

Crossman, E. J. 1996. Taxonomy and distribution. Pages 1-11 *in* J. F. Craig, editor. Pike biology and exploitation. Chapman and Hall, London.

———. 2004. Pike with horns. Pike Anglers' Club of Great Britain. Available: www.pacgb.co.uk/articles/horns.html (November 2006).

Crossman, E. J., and K. Buss. 1965. Hybridization in the family Esocidae. Journal of the Fisheries Research Board of Canada 22:1261-92.

Crossman, E. J., and J. M. Casselman. 1987. An annotated bibliography of the pike, *Esox lucius* (Osteichthyes: Salmoniformes). A Life Sciences Miscellaneous Publication, Royal Ontario Museum, Toronto.

Davis, R. A., and D. H. Schupp. 1987. Comparisons of catches by standard lake survey nets with catches by modified nets. Minnesota Department of Natural Resources, Section of Fisheries Investigational Report 391, St. Paul.

Debates, T. J., C. P. Paukert, and D. W. Willis. 2003. Fish community responses to the establishment of a piscivore, northern pike *(Esox lucius)* in a Nebraska sandhill lake. Journal of Freshwater Ecology 18:353-59.

Diana, J. S. 1979. The feeding pattern and daily ration of a top carnivore, the northern pike *(Esox lucius)*. Canadian Journal of Zoology 57:2121-27.

———. 1980. Diel activity pattern and swimming speeds of northern pike *(Esox lucius)* in Lac Ste. Anne, Alberta. Canadian Journal of Fisheries and Aquatic Sciences 37:1454-58.

———. 1983a. An energy budget for northern pike *(Esox lucius)*. Canadian Journal of Zoology 61:1968-75.

———. 1983b. Growth, maturation, and production of northern pike in three Michigan lakes. Transactions of the American Fisheries Society 112:38-46.

Malette, M. D., and G. E. Morgan. 2005. Provincial summary of northern pike life history characteristics based on Ontario's fall walleye index netting (FWIN) program 1993 to 2002. Publication of the Cooperative Freshwater Ecology Unit, Department of Biology, Laurentian University, Sudbury, Ontario.

Maloney, J., and D. Schupp. 1977. Use of winter-rescue northern pike in maintenance stocking. Minnesota Department of Natural Resources, Section of Fisheries Investigational Report 345, St. Paul.

Mann, R. H. K. 1980. The numbers and production of pike *(Esox lucius)* in two Dorset rivers. Journal of Animal Ecology 49:899-915.

———. 1982. The annual food consumption and prey preferences of pike *(Esox lucius)* in the River Frome, Dorset. Journal of Animal Ecology 51:81-95.

Margenau, T. L. 1987. Vulnerability of radio-tagged northern pike to angling. North American Journal of Fisheries Management 7:158-59.

———. 1995. Stunted northern pike: a case history of community manipulations and field transfer. Wisconsin Department of Natural Resources Research Report 169, Madison.

Margenau, T. L., S. P. AveLallemant, D. Giehtbrock, and S. T. Schram. 2008. Ecology and management of northern pike in Wisconsin. Hydrobiologia 601:111-23.

Margenau, T. L., S. J. Gilbert, and G. R. Hatzenbeler. 2003. Angler catch and harvest of northern pike in northern Wisconsin lakes. North American Journal of Fisheries Management 23:307-12.

Margenau, T. L., S. V. Marcquenski, P. W. Rasmussen, and E. MacConnell. 1995. Prevalence of blue spot disease (esocid herpesvirus-1) on northern pike and muskellunge in Wisconsin. Journal of Aquatic Animal Health 7:29-33.

Margenau, T. L., P. W. Rasmussen, and J. M. Kampa. 1998. Factors affecting growth of northern pike in small northern Wisconsin lakes. North American Journal of Fisheries Management 18:625-39.

Mauk, W. L., and D. W. Coble. 1971. Vulnerability of some fishes to northern pike *(Esox lucius)* predation. Journal of the Fisheries Research Board of Canada 28:957-69.

McCarraher, D. B., and R. E. Thomas. 1972. Ecological significance of vegetation to northern pike, *Esox lucius*, spawning. Transactions of the American Fisheries Society 101:560-63.

McCombie, A. M., and F. E. J. Fry. 1960. Selectivity of gill nets for lake whitefish, *Coregonus clupeaformis*. Transactions of the American Fisheries Society 89:176-84.

McInerny, M. C., and T. K. Cross. 2005. Comparison of day electrofishing, night electrofishing, and trap netting for sampling inshore fish in Minnesota lakes. Minnesota Department of Natural Resources, Section of Fisheries Special Publication 161, St. Paul.

MNDNR (Minnesota Department of Natural Resources). 1993. Manual of instructions for lake survey. Minnesota Department of Natural Resources, Section of Fisheries Special Publication 147, St. Paul.

Miller, L. M., L. Kallemeyn, and W. Senanan. 2001. Spawning-site and natal-site fidelity by northern pike in a large lake: mark-recapture and genetic evidence. Transactions of the American Fisheries Society 130:307-16.

Miller, L. M., and A. R. Kapuscinski. 1996. Microsatellite DNA markers reveal new levels of genetic variation in northern pike. Transactions of the American Fisheries Society 125:971-77.

Miller, L. M., and W. Senanan. 2003. A review of northern pike population genetics research and its implications for management. North American Journal of Fisheries Management 23:297-306.

Miller, R. B., and R. C. Thomas. 1957. Alberta's "pothole" trout fisheries. Transactions of the American Fisheries Society 86(1956):261-68.

Mingelbier, M., P. Brodeur, and J. Morin. 2008. Spatially explicit model predicting the spawning habitat and early stage mortality of northern pike *(Esox lucius)* in a large system: the St. Lawrence River between 1960 and 2000. Hydrobiologia 601:55-69.

Moen, T. 1962. Silver northerns in northwest Iowa: we have them too! Iowa Conservationist 21(6):48.

Moen, T., and D. Henegar. 1971. Movement and recovery of tagged northern pike in Lake Oahe, South and North Dakota, 1964-68. Pages 85-93 *in* G. E. Hall, editor. Reservoir fisheries and limnology. American Fisheries Society, Special Publication 8, Bethesda, Md.

Monson, B. A. 2009. Trend reversal of mercury concentrations in piscivorous fish from Minnesota lakes: 1982-2006. Environmental Science and Technology 43: 1750-55.

Morrow, J. V., Jr., G. L. Miller, and K. J. Killgore. 1997. Density, size, and foods of larval northern pike in natural and artificial wetlands. North American Journal of Fisheries Management 17:210-14.

Mosindy, T. E., W. T. Momot, and P. J. Colby. 1987. Impact of angling on the production and yield of mature walleyes and northern pike in a small boreal lake in Ontario. North American Journal of Fisheries Management 7:493-501.

Moyle, J. B. 1949. Fish-population concepts and management of Minnesota lakes for sport fishing. Transactions of the North American Wildlife Conference 14:283-94.

———. 1950. Gill nets for sampling fish populations in Minnesota waters. Transactions of the American Fisheries Society 79(1949):195-204.

———. 1956. Relationships between the chemistry of Minnesota surface waters and wildlife management. Journal of Wildlife Management 20:303-20.

Moyle, J. B., and W. D. Clothier. 1959. Effects of management and winter oxygen levels on the fish population of a prairie lake. Transactions of the American Fisheries Society 88:178-85.

Moyle, J. B., J. H. Kuehn, and C. R. Burrows. 1950. Fish-population and catch data from Minnesota lakes. Transactions of the American Fisheries Society 78(1948):163-75.

Moyle, J. B., and R. Lound. 1960. Confidence limits associated with means and medians of series of net catches. Transactions of the American Fisheries Society 89:53-58.

Muhlfeld, C. C., D. H. Bennett, R. K. Steinhorst, B. Marotz, and M. Boyer. 2008. Using bioenergetics modeling to estimate consumption of native juvenile salmonids by nonnative northern pike in the Upper Flathead River system, Montana. North American Journal of Fisheries Management 28:636-48.

Murray, B. A., J. M. Farrell, K. L. Schulz, and M. A. Teece. 2008. The effect of egg size and nutrient content on larval performance: implications to protracted spawning in northern pike (*Esox lucius* Linnaeus). Hydrobiologia 601:71-82.

Nelson, W. R. 1978. Implications of water management in Lake Oahe for spawning success of coolwater fishes. Pages 154-58 in R. L. Kendall, editor. Selected coolwater fishes of North America. American Fisheries Society, Special Publication 11, Bethesda, Md.

Neumann, R. M., and D. W. Willis. 1994. Length distributions of northern pike caught in five gill net mesh sizes. Prairie Naturalist 26:11-13.

Nielsen, L. A. 1992. Methods of marking fish and shellfish. American Fisheries Society, Special Publication 23, Bethesda, Md.

Nilsson, P. A. 2006. Avoid your neighbours: size-determined spatial distribution patterns among northern pike individuals. Oikos 113:251-58.

Nursall, J. R. 1973. Some behavioral interactions of spottail shiners *(Notropis hudsonius)*, yellow perch *(Perca flavescens)*, and northern pike *(Esox lucius)*. Journal of the Fisheries Research Board of Canada 30:1161-78.

Olson, D. E., and P. K. Cunningham. 1989. Sport-fisheries trends shown by an annual Minnesota fishing contest over a 58-year period. North American Journal of Fisheries Management 9:287-97.

Orabutt, D. E., Jr. 2006. Northern pike in selected Colorado trout reservoirs. Master's thesis. Colorado State University, Fort Collins.

Otto, C. 1979. The effects on a pike (*Esox lucius* L.) population of intensive fishing in a south Swedish lake. Journal of Fish Biology 15:461-68.

Papas, T. S., J. E. Dahlberg, and R. A. Sonstegard. 1976. Type C virus in lymphosarcoma in northern pike *(Esox lucius)*. Nature 261:506-8.

Paukert, C. P., J. A. Klammer, R. B. Pierce, and T. D. Simonson. 2001. An overview of northern pike regulations in North America. Fisheries 26(6):6-13.

Paukert, C. P., W. Stancill, T. J. DeBates, and D. W. Willis. 2003. Predatory effects of northern pike and largemouth bass: bioenergetic modeling and ten years of fish community sampling. Journal of Freshwater Ecology 18:13-24.

Peake, S. 2004. Effect of approach velocity on impingement of juvenile northern pike at water intake screens. North American Journal of Fisheries Management 24:390-96.

———. 2008. Behavior and passage performance of northern pike, walleyes, and white suckers in an experimental raceway. North American Journal of Fisheries Management 28:321-27.

Peterson, J., M. Taylor, and A. Hanson. 1980. Leslie population estimate for a large lake. Transactions of the American Fisheries Society 109:329-31.

Pierce, R. B. 1997. Variable catchability and bias in population estimates for northern pike. Transactions of the American Fisheries Society 126:658-64.

———. 1998. Northern pike spearing: a staff update. Minnesota Department of Natural Resources, Section of Fisheries Staff Report 55, St. Paul.

———. 2004. Oviduct insertion of radio transmitters as a means of locating northern pike spawning habitat. North American Journal of Fisheries Management 24:244-48.

———. 2010a. Long-term evaluations of length limit regulations for northern pike in Minnesota. North American Journal of Fisheries Management 30:412-32.

———. 2010b. Long-term evaluations of northern pike experimental regulations in Minnesota lakes. Minnesota Department of Natural Resources, Fisheries Management Section Investigational Report 556, St. Paul.

Pierce, R. B., and M. F. Cook. 2000. Recreational darkhouse spearing for northern pike in Minnesota: historical changes in effort and harvest and comparisons with angling. North American Journal of Fisheries Management 20:239–44.

Pierce, R. B., L. W. Kallemeyn, and P. J. Talmage. 2007. Light trap sampling of juvenile northern pike in wetlands affected by water level regulation. Minnesota Department of Natural Resources, Fisheries Management Section Investigational Report 550, St. Paul.

Pierce, R. B., S. Shroyer, B. Pittman, D. E. Logsdon, and T. D. Kolander. 2006. Catchability of larval and juvenile northern pike in Quatrefoil light traps. North American Journal of Fisheries Management 26:908–15.

Pierce, R. B., and C. M. Tomcko. 1993. Tag loss and handling mortality for northern pike marked with plastic anchor tags. North American Journal of Fisheries Management 13:613–15.

———. 1997. Initial effects of slot length limits for northern pike in five north-central Minnesota lakes. Minnesota Department of Natural Resources, Section of Fisheries Investigational Report 454, St. Paul.

———. 1998. Angler noncompliance with slot length limits for northern pike in five small Minnesota lakes. North American Journal of Fisheries Management 18:720–24.

———. 2003a. Interrelationships among production, density, growth, and mortality of northern pike in seven north-central Minnesota lakes. Transactions of the American Fisheries Society 132:143–53.

———. 2003b. Variation in gill-net and angling catchability with changing density of northern pike in a small Minnesota lake. Transactions of the American Fisheries Society 132:771–79.

———. 2005. Density and biomass of native northern pike populations in relation to basin-scale characteristics of north-central Minnesota lakes. Transactions of the American Fisheries Society 134:231–41.

Pierce, R. B., C. M. Tomcko, and M. T. Drake. 2003. Population dynamics, trophic interactions, and production of northern pike in a shallow bog lake and their effects on simulated regulation strategies. North American Journal of Fisheries Management 23:323–30.

Pierce, R. B., C. M. Tomcko, and T. D. Kolander. 1994. Indirect and direct estimates of gill-net size selectivity for northern pike. North American Journal of Fisheries Management 14:170–77.

Pierce, R. B., C. M. Tomcko, and T. L. Margenau. 2003. Density dependence in growth and size structure of northern pike populations. North American Journal of Fisheries Management 23:331–39.

Pierce, R. B., C. M. Tomcko, D. L. Pereira, and D. F. Staples. 2010. Differing catchability among lakes: the influences of lake basin morphology and other factors on gill-net catchability of northern pike. Transactions of the American Fisheries Society 139:1109–20.

Pierce, R. B., C. M. Tomcko, and D. H. Schupp. 1995. Exploitation of northern pike in seven small north-central Minnesota lakes. North American Journal of Fisheries Management 15:601-9.

Pierce, R. B., J. A. Younk, and C. M. Tomcko. 2005. Expulsion of miniature radio transmitters along with eggs of northern pike and muskellunge: a new method for locating critical spawning habitat. Minnesota Department of Natural Resources, Section of Fisheries Investigational Report 522, St. Paul.

———. 2007. Expulsion of miniature radio transmitters along with eggs of muskellunge and northern pike: a new method for locating critical spawning habitat. Environmental Biology of Fishes 79:99-109.

Prejs, A., A. Martyniak, S. Boron, P. Hliwa, and P. Koperski. 1994. Food web manipulation in a small, eutrophic Lake Wirbel, Poland: effect of stocking with juvenile pike on planktivorous fish. Hydrobiologia 275/276:65-70.

Priegel, G. R., and D. C. Krohn. 1975. Characteristics of a northern pike spawning population. Wisconsin Department of Natural Resources Technical Bulletin 86, Madison.

Pullen, M. M., J. C. Schlotthauer, and J. O. Hanson. 1981. Fish tapeworm and human health. University of Minnesota Agricultural Extension Service, Veterinary Science Fact Sheet 28-1981.

Raat, A. J. P. 1988. Synopsis of biological data on the northern pike *Esox lucius* Linnaeus, 1758. Food and Agriculture Organization of the United Nations, FAO Fisheries Synopsis No. 30 Rev. 2, Rome.

Radomski, P. 2006. Historical changes in abundance of floating-leaf and emergent vegetation in Minnesota lakes. North American Journal of Fisheries Management 26:932-40.

Radomski, P., and T. J. Goeman. 2001. Consequences of human lakeshore development on emergent and floating-leaf vegetation abundance. North American Journal of Fisheries Management 21:46-61.

Rask, M., and L. Arvola. 1985. The biomass and production of pike, perch and whitefish in two small lakes in southern Finland. Annales Zoologici Fennici 22:129-36.

Reznick, D. N., and C. K. Ghalambor. 2005. Can commercial fishing cause evolution? Answers from guppies *(Poecilia reticulata)*. Canadian Journal of Fisheries and Aquatic Sciences 62:791-801.

Rich, B. A. 1992. Population dynamics, food habits, movement and habitat use of northern pike in the Coeur d'Alene Lake system, Idaho. Idaho Department of Fish and Game, Project F-73-R-14 Completion Report, Subproject VI, Study 3, Boise.

Ricker, W. E. 1975. Computation and interpretation of biological statistics of fish populations. Fisheries Research Board of Canada Bulletin 191, Ottawa.

Robinson, C. L. K. 1989. Laboratory survival of four prey in the presence of northern pike. Canadian Journal of Zoology 67:418-20.

Ross, M. J., and J. D. Winter. 1981. Winter movements of four fish species near a thermal plume in northern Minnesota. Transactions of the American Fisheries Society 110:14-18.

Rudenko, G. P. 1971. Biomass and abundance in a roach-perch lake. Journal of Ichthyology 11:524-35.

Sammons, S. M., C. G. Scalet, and R. M. Neumann. 1994. Seasonal and size-related changes in the diet of northern pike from a shallow prairie lake. Journal of Freshwater Ecology 9:321-29.

Scheirer, J. W. 1988. Angling characteristics and vital statistics of fish populations in Long Lake, Fond Du Lac County, Wisconsin. Master's thesis. University of Wisconsin, Stevens Point.

Scheirer, J. W., and D. W. Coble. 1991. Effects of Floy FD-67 anchor tags on growth and condition of northern pike. North American Journal of Fisheries Management 11:369-73.

Schramm, S. T. 1983. Population characteristics of northern pike in a Lake Superior estuary. Wisconsin Department of Natural Resources, Fish Management Report 115, Madison.

Schulze, T., U. Baade, H. Dorner, R. Eckmann, S. S. Haertel-Borer, F. Holker, and T. Mehner. 2006. Response of the residential piscivorous fish community to introduction of a new predator type in a mesotrophic lake. Canadian Journal of Fisheries and Aquatic Sciences 63:2202-12.

Schupp, D. H. 1981. A review of the status of northern pike in Minnesota. Minnesota Department of Natural Resources, Section of Fisheries Staff Report 25, St. Paul.

———. 1992. An ecological classification of Minnesota lakes with associated fish communities. Minnesota Department of Natural Resources, Section of Fisheries Investigational Report 417, St. Paul.

Schwarz, C. J. 2006. Analysis of the mark-recapture studies for northern pike in Mille Lacs, Minnesota. Report for contract A82526 for the Minnesota Department of Natural Resources.

Scidmore, W. J. 1964. Use of yearling northern pike in the management of Minnesota fish lakes. Minnesota Department of Conservation, Division of Game and Fish Investigational Report 276, St. Paul.

Scott, W. B., and E. J. Crossman. 1973. Freshwater fishes of Canada. Fisheries Research Board of Canada Bulletin 184, Ottawa.

Seaburg, K. G., and J. B. Moyle. 1964. Feeding habits, digestive rates, and growth of some Minnesota warmwater fishes. Transactions of the American Fisheries Society 93:269-85.

Seber, G. A. F. 1982. The estimation of animal abundance and related parameters, 2nd edition. MacMillan, New York.

Seeb, J. E., L. W. Seeb, D. W. Oates, and F. M. Utter. 1987. Genetic variation and postglacial dispersal of populations of northern pike *(Esox lucius)* in North America. Canadian Journal of Fisheries and Aquatic Sciences 44:556-61.

Senanan, W., and A. R. Kapuscinski. 2000. Genetic relationships among populations of northern pike *(Esox lucius)*. Canadian Journal of Fisheries and Aquatic Sciences 57:391-404.

Sheehan, R. J., W. M. Lewis, L. R. Bodensteiner, M. Schmidt, E. Sandberg, and G. Conover. 1994. Winter habitat requirements and overwintering of riverine fishes. Southern Illinois University, Fisheries Research Laboratory Final Performance Report F-79-R-6, Carbondale.

Siefert, R. E., W. A. Spoor, and R. F. Syrett. 1973. Effects of reduced oxygen concentrations on northern pike *(Esox lucius)* embryos and larvae. Journal of the Fisheries Research Board of Canada 30:849-52.

Skov, C., and S. Berg. 1999. Utilization of natural and artificial habitats by YOY pike in a biomanipulated lake. Hydrobiologia 408/409:115-22.

Skov, C., L. Jacobsen, and S. Berg. 2003. Post-stocking survival of 0+ year pike in ponds as a function of water transparency, habitat complexity, prey availability and size heterogeneity. Journal of Fish Biology 62:311-22.

Skov, C., and P. A. Nilsson. 2007. Evaluating stocking of YOY pike *Esox lucius* as a tool in the restoration of shallow lakes. Freshwater Biology 52:1834-45.

Snow, H. E. 1974. Effects of stocking northern pike in Murphy Flowage. Wisconsin Department of Natural Resources Technical Bulletin 79, Madison.

———. 1978. Responses of northern pike to exploitation in Murphy Flowage, Wisconsin. Pages 320-27 in R. L. Kendall, editor. Selected coolwater fishes of North America. American Fisheries Society, Special Publication 11, Bethesda, Md.

Snow, H. E., and T. D. Beard. 1972. A ten-year study of native northern pike in Bucks Lake, Wisconsin including evaluation of an 18.0-inch size limit. Wisconsin Department of Natural Resources, Technical Bulletin 56, Madison.

Søndergaard, M., E. Jeppesen, J. P. Jensen, and T. Lauridsen. 2000. Lake restoration in Denmark. Lakes and Reservoirs: Research and Management 5:151-59.

Sonstegard, R. A., and J. G. Hnath. 1978. Lymphosarcoma in muskellunge and northern pike: guidelines for disease control. Pages 235-37 in R. L. Kendall, editor. Selected coolwater fishes of North America. American Fisheries Society, Special Publication 11, Bethesda, Md.

Sorensen, J. A., G. E. Glass, K. W. Schmidt, J. K. Huber, and G. R. Rapp Jr. 1990. Airborne mercury deposition and watershed characteristics in relation to mercury concentrations in water, sediments, plankton, and fish of eighty northern Minnesota lakes. Environmental Science and Technology 24:1716-27.

Soupir, C. A., M. L. Brown, and L. W. Kallemeyn. 2000. Trophic ecology of largemouth bass and northern pike in allopatric and sympatric assemblages in northern boreal lakes. Canadian Journal of Zoology 78:1759-66.

Spens, J., and J. P. Ball. 2008. Salmonid or nonsalmonid lakes: predicting the fate of northern boreal fish communities with hierarchical filters relating to a keystone piscivore. Canadian Journal of Fisheries and Aquatic Sciences 65:1945-55.

Stefan, H. G., M. Hondzo, X. Fang, J. E. Eaton, and J. H. McCormick. 1996. Simulated long-term temperature and dissolved oxygen characteristics of lakes in the north-central United States and associated fish habitat limits. Limnological and Oceanographic Research 41:1124-35.

Tomcko, C. M. 1997. A review of northern pike *Esox lucius* hooking mortality. Minnesota Department of Natural Resources, Section of Fisheries, Fish Management Report 32, St. Paul.

Trautman, M. B., and C. L. Hubbs. 1935. When do pike shed their teeth? Transactions of the American Fisheries Society 65:261-66.

Treasurer, J. W., R. Owen, and E. Bowers. 1992. The population dynamics of pike, *Esox lucius*, and perch, *Perca fluviatilis*, in a simple predator-prey system. Environmental Biology of Fishes 34:65-78.

Venturelli, P. A., and W. M. Tonn. 2006. Diet and growth of northern pike in the absence of prey fishes: initial consequences for persisting in disturbance-prone lakes. Transactions of the American Fisheries Society 135:1512-22.

Verta, M. 1990. Changes in fish mercury concentrations in an intensively fished lake. Canadian Journal of Fisheries and Aquatic Sciences 47:1888-97.

Vessel, M. F., and S. Eddy. 1941. A preliminary study of the egg production of certain Minnesota fishes. Minnesota Department of Conservation Fisheries Research Investigational Report 26, St. Paul.

Wahl, D. H., and R. A. Stein. 1988. Selective predation by three esocids: the role of prey behavior and morphology. Transactions of the American Fisheries Society 117:142-51.

Walker, J. R., L. Foote, and M. G. Sullivan. 2007. Effectiveness of enforcement to deter illegal angling harvest of northern pike in Alberta. North American Journal of Fisheries Management 27:1369-77.

Waters, T. F. 1960. The development of population estimate procedures in small trout lakes. Transactions of the American Fisheries Society 89:287-94.

Webb, P. W. 1984. Body and fin form and strike tactics of four teleost predators attacking fathead minnow *(Pimephales promelas)* prey. Canadian Journal of Fisheries and Aquatic Sciences 41:157-65.

Webb, P. W., and J. M. Skadsen. 1980. Strike tactics of *Esox*. Canadian Journal of Zoology 58:1462-69.

Weithman, A. S., and R. O. Anderson. 1978. Angling vulnerability of Esocidae. Proceedings of the Annual Conference, Southeastern Association of Fish and Wildlife Agencies 30:99-102.

Wesloh, M. L., and D. E. Olson. 1962. The growth and harvest of stocked yearling northern pike, *Esox lucius* Linnaeus, in a Minnesota walleye lake. Minnesota Department of Conservation, Division of Game and Fish Investigational Report 242, St. Paul.

Wiener, J. G., B. C. Knights, M. B. Sandheinrich, J. D. Jeremiason, M. E. Brigham, D. R. Engstrom, L. G. Woodruff, W. F. Cannon, and S. J. Balogh. 2006. Mercury in soils, lakes, and fish in Voyageurs National Park (Minnesota): importance of atmospheric deposition and ecosystem factors. Environmental Science and Technology 40:6261-68.

Williams, J. E. 1955. Determination of age from the scales of northern pike (*Esox lucius* L.). Doctoral dissertation. University of Michigan, Ann Arbor.

Winfield, I. J., J. B. James, and J. M. Fletcher. 2008. Northern pike *(Esox lucius)* in a warming lake: changes in population size and individual condition in relation to prey abundance. Hydrobiologia 601:29-40.

Woods, D. E. 1963. Contribution to the fishery of a northern pike year class of known strength, 1962. Minnesota Department of Conservation, Division of Game and Fish Investigational Report 263, St. Paul.

Wright, R. M. 1990. The population biology of pike, *Esox lucius* L., in two gravel pit lakes, with special reference to early life history. Journal of Fish Biology 36:215-29.

Wright, R. M., and E. A. Shoesmith. 1988. The reproductive success of pike, *Esox lucius*: aspects of fecundity, egg density and survival. Journal of Fish Biology 33:623-36.

removals, 51-52, 57-58, 63-65, 73, 146, 165

scales of pike, 41-45
scientific name, 1
sexes, 21-23, 29, 45-46, 58
sexual maturity, 21-22
silver pike, 2-5
size structure of populations, 14, 40-41, 52-54, 73, 88, 92, 95-97, 100, 103-12, 143, 175
social issues and pike management, 78-81, 91, 110-11, 175-77
spawning: behavior, 24-29; egg laying, 24-26; habitat, 24-28; movements, 24-28; site fidelity, 26-28. *See also* managed spawning and rearing areas
spearing. *See* darkhouse spearing
special regulations. *See* experimental regulations
stocking of pike: historical records, 112, 118-20, 124-25, 132-34; issues, 63-64, 66-67, 91-95, 126
stunted fish, 51, 62, 92, 95-96, 111-12
swimming speeds and flowing water, 20-21

tagging pike, 55, 161-64; Dennison tags, 161-63; Floy tags, 161-64; handling mortality, 163; tag loss, 162-64
teeth of pike, 68-69
telemetry: locating spawning sites, 170-73; oviduct implantation, 168-73; radio and ultrasonic, 12, 17, 19-20, 166-73; uses, 166-67, 170, 173
trap netting (spring ice-out), 151-58; mortality in trap nets, 154; net construction/dimensions, 153-54

vulnerability to angling, 59-60, 71-73

water temperature, 8-11, 15-16, 24, 30-32, 38, 43, 46-47, 50, 118, 154, 167, 179
winter rescue: conservation issues, 93-95, 126-27, 134; methods/techniques, 93, 97, 118-23; pike yields, 119, 123-25

year-class strength, 32, 38-40, 68, 132-33
yields of pike, 55, 58-59, 81, 93, 123-24, 132-34

Rodney B. Pierce is a fisheries research biologist with the Minnesota Department of Natural Resources.